Jorge Alfredo Gonzélez Pérez
Felicidad Ndafahamba Alberto Cabral

Proposal for a climate variable information system

Jorge Alfredo Gonzélez Pérez
Felicidad Ndafahamba Alberto Cabral

Proposal for a climate variable information system

Proposal for a climate variable information system for agriculture in the municipality of Cuanhama, Cunene. Angola

ScienciaScripts

Imprint

Cover image: www.ingimage.com

This book is a translation from the original published under ISBN 978-620-0-01022-3.

Publisher:
Sciencia Scripts
is a trademark of
Dodo Books Indian Ocean Ltd. and OmniScriptum S.R.L publishing group

120 High Road, East Finchley, London, N2 9ED, United Kingdom
Str. Armeneasca 28/1, office 1, Chisinau MD-2012, Republic of Moldova, Europe
Managing Directors: Ieva Konstantinova, Victoria Ursu
info@omniscriptum.com

Printed at: see last page
ISBN: 978-620-8-51866-0

THANKS.

- *To Almighty God for the gift of life and health, for the strength, wisdom and ability to make it this far.*
- *To my teachers who dedicated their time and gave everything to make this moment happen.*
- *I would like to thank my dear tutor JORGE ALFREDO GONZÁLEZ PÉREZ for his patience and dedication, thank you.*
- *I thank three very important people in my life who were present from the beginning for their advice and financial support, my mother Gizela J. Dembo, my uncle Estefanio, you are my second parents.*
- *I thank my love Nelson K. Hach for his unconditional support and companionship, motivation and support to make all this possible, thank you love.*
- *To my dear colleagues for the companionship and the battles we went through.*

Thank you all for your support.

DEDICATION

I dedicate this monograph to my dear family, especially to my mother.ely "Augusta Inés Augusta" for all the prayers and support in the difficult times we went through, even so we are still standing with little that you worked hard for your daughter to have a degree today, thank you, I love you.

My princess Azucena Tuyoleni Hach whom I love very much and may she be proud of your mother and may you follow her as an example, I love you.

SUMMARY

This work is carried out in the municipality of Cuanhama, Cunene Province, where the historical behaviour of some climatic variables such as maximum and minimum temperature, relative humidity and precipitation was analysed during a historical period of ten years (2013-2022) and compared with the year 2023. This research was carried out with information received from the Ministry of Telecommunications, Information Technologies and Social Communication. National Institute of Meteorology and Geophysics Provincial Meteorological Department of Cunene, where data on maximum and minimum temperature, relative humidity and precipitation were obtained for 11 years. The average maximum temperature of the year analysed (2023) had a value of 31.58 ºC, being lower than the historical average to which I put an annual average value of 34.7 ºC, the minimum humidity in the year analysed was 17.2 ºC higher than the historical average (13.3 ºC), the relative humidity of the historical average 52.7% was higher than that of the year analysed 34.0%, the rainfall in both cases was low although the historical average (37.5 mm) was higher than that of the year 2023 (32.6 millimetres). An agro-climatic bulletin is produced that provides detailed information on weather conditions in the area, as well as useful elements and information for producers to enable them to make favourable decisions in their production. GRAMOensure that producers are updated in relation to them, having information that helps them to make decisions regarding the types of production to sow and develop in their gardens, so as to minimise economic damage to them.

Key words: variability, temperature, humidity, precipitation,

INDEX

I. INTRODUCTION

Timely knowledge and information, with continuous monitoring of environmental variables, are essential elements for profit-oriented decision-making for sustainable development. Therefore, all information on vulnerability, impacts and adaptation to climate change constitutes a positive contribution to the task in which the countries of Latin America and the Caribbean are involved (IPCC, 2007a).

During the 20th century, increases in temperature, increases and decreases in precipitation and changes in the occurrence of extreme events were reported in several areas. Changes in extreme events included positive trends in warm nights, heavy rainfall and consecutive dry days. In Latin American and Caribbean countries, there is ample evidence of increases in the occurrence of these climate events and climate change, such as the continuing episodes of droughts and floods, Hurricane Catarina in the South Atlantic and the record number of hurricanes that occurred in 2005 in the Caribbean (IPCC, 2007a).

Negative impacts of these changes included glacier retreat, increased frequency of flooding, increased morbidity and mortality, increased forest fires, loss of biodiversity, increased plant diseases, reduced dairy cattle production and problems with hydropower generation (IPCC, 2007a).

Africa has not been exempt from these events, and the last decades have seen an increase in climate variability. Centelha et al., (1997) observed in Africa an increase in temperatures and a significant increase in the number or intensity of droughts, which not only doubled their frequency of occurrence, but also recorded a considerable increase in the number of extreme cases.

Segundo Sánchez-Santilán et al., (2014), In the last two years rainfall has had a positive impact on agriculture and in all areas of society, this last year is the only one that exceeded the annual average, in any case, we are not exempt from climate change and from suffering a severe drought again, it is necessary to be prepared as the risk is present, since an increase in evaporation and temperatures (between 0.6 and 3.5 degrees Celsius) is reported.

The trend towards increased drought coincides with future projections associated with climate change, which occurred in the country during the period 1961-1990, moderate and severe drought events, which doubled in the period 1931-1960, after 1990 there were significant drought events, with greater incidence in the eastern provinces. The five eastern provinces were subjected to extreme weather events during the years 1998 - 2007, according to studies conducted by the Institute of Physical Planning Sanchez-Santilan et al., (2014), The province of Cunene is one of the most vulnerable to food insecurity, during this period the effect of agricultural drought severely influenced agriculture in the province.

The municipality of Cuanhama in the province of Cunene was not exempt from the aforementioned events, which is why it needs to have sufficient information to enable social actors and decision-makers to make decisions quickly and accurately.

In this sense it is declared as: Scientific Problem: Insufficiency in the information system of climatic variables that prevent the producers of the municipality of Cuanhama in the province of Cunene to take better decisions for the results of their production.

Based on the above, the following hypothesis is presented: "If the information system is improved to allow the integration of the climatological variables of the municipality of Cuanhama, it will be possible to contribute to their efficient use in decision-making in the agricultural sector, depending on local development.

In this research the General Objective is: To propose a ten-year bulletin that allows the integration of climatological variables in the municipality, in order to contribute to the decision-making process of local decision-makers.

is set as specific objectives:

1. To analyse the historical behaviour of the variables temperature, relative humidity and precipitation in the study area in the period 2013-2022.
2. Produce a ten-yearly bulletin containing a system of information on climatic variables favourable to producers who can make decisions on their production.

II. LITERATURE REVIEW

2.1. Climatological variables

International literature includes different methods to determine the influence of climatic variables on drought, both meteorological and agricultural, and among them, the most important and usually considered climatic element is the precipitation variable. Some of them are based on monthly analyses of atmospheric meteorological information. In agricultural applications, one month is a very long period and not suitable for assessing plant water deficit under dry conditions (Villalpando, 2024).

Rainfall can be concentrated in a certain period of the month, thus creating other drier periods in the rest of the month, so studying the influence of agricultural drought, which can be critical in certain phases of plant development, recommended organising the information in periods of ten years (10 days), to increase its accuracy. Serrano et al., (2017).

These authors go on to explain that climatic indices that relate elements of weather and climate to drought are used to quantitatively determine meteorological drought. Many of them associate the precipitation variable with the temperature variable, the evaporation variable and other climatic elements. Agro-climatic indices derived from basic climatological variables generally indicate the temporal and spatial extent and intensification of drought. Any study and development of the agricultural drought index must faithfully and accurately reflect this complex phenomenon.

It must also define the extent to which plants are impaired in their growth and development by the stresses to which they have been subjected by moisture deficit stress and quantify in detail the supply and demand for water in the complex to which I am accustomed. plant atmosphere, therefore, agricultural drought indices are developed primarily based on the variables precipitation, evapotranspiration, evaporation, soil water content and crop development status (Paneque & Habana, 1998).

In the context of agriculture, drought "does not begin when rainfall ceases, except when plant roots can no longer obtain moisture from the soil" and can be defined on the basis of soil moisture and not on some indirect interpretation of rainfall records. Since the productive soil moisture reserve depends on the character of the soil and the crop, agricultural drought occurs when soil moisture in the rhizosphere is at a level that limits crop growth and production (Paneque and Habana, 1998).

According to the World Meteorological Organisation, agricultural drought occurs when the amount of precipitation and its distribution, soil water reserves and evaporation losses combine to cause significant decreases in crop and livestock yields. This results in depressed food production, inadequate grazing conditions, low returns on agricultural labour and investment, reduced availability of wood for fuel, increased potential danger of vegetation fires, increased risk of desertification and the social impact and economic consequences linked to drought, including food supply insecurity (Montenegro and Luján, 2018). (Montenegro and Luján, 2018).

When water availability is lower than the needs of the vegetation to meet its normal growth and development, although water is theoretically available up to the point of permanent wilting, water consumption by plants is progressively reduced from the productive limit (lowest value of soil water content, above which plants can reach maximum yield, in the absence of limiting factors), up to the point of wilting.

In this case, the hydraulic conductivity of the soil does not allow water to be transported quickly enough to the plant roots to meet the transpiration demand. Soil moisture is expressed by the amount of water it contains or also by the tension at which water is retained (Villalpando, 2024)

The first expression is used to study water losses or gains and the second to understand the movement and availability of water for plants. As a result of this balance we obtain, through a spreadsheet, Serrano et al. (2017), created in the development of the model for the calculation of the soil water balance in the rhizosphere, the water availability index for a drainage point that covers the whole country and whose step is 1 km on one side.

The quantitative relationship between this index and precipitation, the productive moisture reserve, the water needs of plants and the water stress to which the vegetation is subjected, are the variables that give rise to the modelling to develop an agrometeorological index that allows determining the moisture conditions formed in the predominant vegetation in the study area and other issues related to the moisture in various aspects related to agriculture, recognized by Gutiérrez and Anay (2019), Gosling and Arnell (2016) and jiménez et al,(2004), as a wetting index.

Continuous monitoring of the moisture index conditions of a location allows detecting the presence of extreme drought and moisture events detrimental to agriculture. The practical application of the model generated by FAO (2010) and Gyoo-Bum et al. (2006), calculating the agricultural drought index and evaluating, in the agrometeorological monitoring system, the development of the processes of this extreme climatic event in the different episodes of agricultural drought from 1996 to date, Chavarría et al. (2020) and Fernando (2020) indicate that this model has described with very good approximation the onset, spatial extent, evolution, reduction of the affected area, weakening, end and absence of agricultural drought.

However, the experience gained in the use of the model made it possible to calibrate the end of drought category, making it correspond to the disappearance of water stress conditions in the vegetation.

Based on the monitoring of dry weather periods evaluated by the modified FAO humidity index (2007) and the scale proposed by Chavarría et al. (2007), and the scale proposed by Chavarría et al., (2020) which allows to evaluate the onset, end and duration of agricultural drought, depending on the water stress conditions affecting the predominant vegetation in the study area, the evolution of agricultural drought is conceptualised in six categories that we list below according to Fernando, (2020):

a) Short dry period.

This category expresses that a dry period occurred in which the vegetation depleted the productive soil moisture reserve and remained in moderate to severe water stress for the next two decades (very dry or severely dry moisture index categories). Its effects may correspond, in terms of plant water stress, to the categories of absolute drought, dry period or partial drought.

b) Moderate dry period.

Corresponds to that dry period that kept the vegetation under moderate or severe water stress for an additional period of one decade to the category described above and whose duration was greater than or equal to three decades after the onset of soil water depletion.

c) Beginning of the agricultural drought.

It comprises that dry period that kept the vegetation under moderate or severe water stress for a period additional to one decade to the category described above and whose duration was greater than or equal to four decades after the onset of soil water depletion. This sufficiently long time period corresponds to the onset of meteorological drought in the Climate Centre's Drought Monitoring System.

d) Continued agricultural drought.

This category indicates the establishment of an agricultural drought. The agricultural drought, in this case, kept the vegetation under moderate or severe water stress for an additional period of one decade to the category already described above and its duration was greater than or equal to five decades after the onset of drought on the soil. depletion.

e) End of the agricultural drought.

It starts with the first decade, between two consecutive decades, where there are wet conditions after a dry period, and there may be a dry period between them that causes slight water stress to the plants (the water content of the soil does not allow the plants to meet their water needs, i.e. the productive moisture of the product is less than the volume of water it stores at the productive limit and is greater than or equal to half the water content it stores when it is at the productive limit, i.e. it does not extract water from the soil easily, nor at high tensions.

Vegetation, under these conditions, almost satisfies its water needs and extracts water from the soil with some limitations, grows with some difficulties and this slightly reduces its growth and biomass production).

f) Absence of agricultural drought.

This is the period when agro-meteorological conditions do not result in dry periods or droughts. In this case, the plants can obtain economically acceptable yields and even reach their maximum yields.

According to research results from the Climate Centre of the Institute of Meteorology, in the last three decades air temperature has increased by 0.6 °C, the occurrence of severe storms accompanied by strong winds and lightning has increased, and the frequency and intensity of droughts and droughts have increasedIDEAM-UNAL (2018).

In Angola's climatic conditions, there are two very different seasonal periods, rainy and dry. In general, the rainy season coincides with the summer. Approximately three quarters of the total annual rainfall accumulates there. The lack of rainfall during this period may characterise the onset or presence of meteorological drought, although no stress conditions occur in the plants, as the vegetation during this period, under dry conditions, is often able to meet its water needs or be very dry. close to this, by consuming the water reserve of the soil.

In the period of low rainfall, vegetation in dry conditions tends not to meet its moisture requirements and stressful conditions for vegetation prevail, although there is no precipitation deficit that causes meteorological drought Inzunza, (2023).

Temperature is another important climate variable with scientific evidence, at global, hemispheric, regional and sub-regional scales, of an increase in surface air temperature over the last decades, and that this increase is related to anthropogenic greenhouse gas emissions. This increase has implications for sustainability because it causes negative impacts on most socio-economic sectors and ecosystems, with consequent effects on biodiversity (UNDP, 2007).

2.2. Climate, its variations and changes

Climate is dynamic and has intrinsic variability. The term variability is used to indicate separations of climate statistics in periods of months, seasons or years, with respect to statistics of different duration referring to the same period (month, season or year) and is measured by these separations known as anomalies. When differences are observed between long-term statistics of climatic elements calculated for different periods, but relating to the same area, it is said that we are in the presence of climate change (CMMAD, 1987).

According to Guzmán-Castellanos et al. (2014), climate is the result of the complex interactions of many components, and its behaviour is highly dynamic and non-linear. In recent decades, increased international attention has focused on the issue of climate variability, as extreme climatic anomalies have occurred in many regions of the planet, including intense and extensive drought processes, severe and devastating rainfall events, extraordinarily warm temperatures. years and many other phenomena that caused significant human and material impacts in many countries, in some cases reaching the classification of disasters of immense proportions.

The best defence strategy in the face of extreme climatic anomalies is risk preparedness, i.e. the practical adoption of proactive or anticipatory countermeasures that take into account the likely recurrence of the phenomenon in question and its characteristics of manifestation, vulnerability of the exposed elements, etc. This avoids surprises and, consequently, reactive or improvised measures, which are often counterproductive and incompatible with the principle of sustainability (Llacza et al., 2016).

Climate has undergone many changes in the past. Geological data from previous millennia and instrumental records indicate that climatic variations have always existed, sometimes to extremes.

There is evidence that climatic changes over the last two million years caused cattle, cattle and livestock to migrate as they moved towards or away from glaciers. Within these major climatic upheavals, other, more rapid fluctuations have been shown to occur on timescales of hundreds of years, and are far more important to humanity because of their immediate consequences (Duval and Campo, 2017).

What is climate change?

A climate change occurs when the temperature of the Earth-atmosphere system, defined as the global mean temperature near the Earth's surface, is determined by the balance between incoming solar radiation (short wavelength) and outgoing radiation, which comprises part of the reflected solar radiation (short wavelength) and terrestrial infrared radiation (broad wavelength). On long time scales (decades or longer) there is a balance between incoming and outgoing energy (Villalpando, 2024).

This means that the temperature of the Earth-atmosphere system remains roughly constant on these timescales. If this equilibrium is changed (the imbalance called radiative forcing), then the system responds to the radiative forcing and attempts to restore the equilibrium by changing its temperature (uribeyAlberto, 2017; Vinet and Zhedanov, 2010).

Processes that produce radiative forcing are known as external forcing factors, which can be natural or anthropogenic. An example of natural external forcing may be a change in the energy emitted by the sun, while an example of anthropogenic (man-made) external forcing is the enhancement of the greenhouse effect.

The greenhouse effect is a natural process that causes warming of the earth's surface and lower atmosphere. It is caused by the presence in the atmosphere of gases that have the ability to absorb and re-emit terrestrial radiation.

Water vapour, carbon dioxide, methane, nitrous oxide, chlorofluorocarbons and tropospheric ozone absorb most of the radiation emitted by the earth's surface, so that the presence of these gases becomes a "trap" for terrestrial radiation, retaining most of the heat carried by this radiation and warming the earth's surface and lower atmosphere.

These gases act in a similar way to the glass in a greenhouse, which is why we call this phenomenon the greenhouse effect and the gases that produce it the greenhouse gases.

Thanks to this effect, the temperature of the planet is 15 ºC and not -18 ºC, i.e. 33 ºC higher than the greenhouse gas that would not be present in the atmosphere, which makes the existence of current forms of life on planet Earth difficult. (Users, 2012).

Until recently, climate change was associated only with external natural forces. Today, however, it is clear that, due to the development patterns used, humans can change the Earth's climate, leading to a strengthening of the greenhouse effect. Large amounts of carbon dioxide are emitted into the atmosphere due to the combustion of fossil fuels, and it is estimated that after the industrial revolution of the 18th century, carbon dioxide emissions increased exponentially. Significant amounts of the same gas were also released into the atmosphere as a result of deforestation and other human activities (Clemente *et al.,* 2016)

2.3. The phenomenon of drought

Drought, in the first instance, is a climatic phenomenon caused by a reduction in precipitation that manifests itself slowly, affecting people, economic activities and the environment in general. In the broadest sense, a drought event occurs when the availability of freshwater, from rainfall, rivers, ponds, underground shells (all sources), is continuously below normal values (Wlemente *et al.,* 2016).

SegundoVillalpando (2024), The phenomenon of drought can be evaluated from different points of view, either by the climatic conditions involved in it, such as rainfall, temperature, evaporation, etc., or by its agricultural, hydrological or economic consequences. Four main types of drought can be identified: meteorological, agricultural, hydrological and social or economic. The first one is related to climatic conditions and the last three to the consequences.

a) *Meteorological drought:* Occurs when precipitation is much lower than expected over a large area and for a prolonged period of time.
b) *Hydrological drought:* Occurs when there is a continuous deficit in surface runoff reaching a level below normal conditions or when the groundwater level decreases.
c) *Agricultural drought:* when the amount of precipitation and its distribution, soil water reserves and evaporation losses combine to cause significant decreases in crop and livestock yields.

d) *Social or economic drought:* This is attributed to natural but also social effects. It is represented by water scarcity induced by an imbalance between water supply and demand.

In recent decades, the phenomenon of drought has been one of the most aggravated climate risks globally. Frequent and persistent droughts have occurred in both developed and developing regions, causing severe impacts on their economies, societies and the environment in general. Among natural disasters, drought is considered to be one of those that affect the greatest number of people (UNEP, 2008).

Second (Clemente *et al.,* 2016), each territory, according to the behaviour of this phenomenon, will develop an adaptation policy based on a structured process that allows for the development of adaptation strategies, policies and measures to ensure human development in the face of climate variability and change. Linking adaptation to climate change with sustainable development and global environmental issues. Allowing for complementing existing policies in developing countries, including project assessment, development and monitoring processes.

Based on four basic principles:

a) Adaptation policies and measures need to be assessed in the context of development.
b) Adaptation to short-term climate variability and extreme events is explicitly included as a step towards reducing vulnerability to long-term climate change.
c) Adaptation occurs at different levels of society, including the local level.
d) The adaptation strategy and the processes through which it is implemented are equally important.

2.4. Climate change science

Since the beginning of the planet, the interaction of the sun, moon, atmosphere, hydrosphere, land cover and other planetary components has been constant and dynamic. The planet's climate has never been static; on the contrary, the interactions between all its components have given rise to different forms and periods. Changes in climatic variables did not allow or limit the biological existence of various species (Maderey and Jiménez, 2001, Sófocleo, 2002 and Núñez and Núñez, 2011).

Throughout history, climate has been one of the determining factors in the development of physical-natural systems and the evolution of societies. Climate configuration has led to the accessibility of fresh water, biological diversification, the habitability of certain regions and agricultural cycles. The configuration of fate is sensitive to natural and human modifications. These changes lead to imbalances in the interaction of elements and systems that feed back into the planet's climate, causing regional and global changes in physical-natural systems. Since climate is variable, significant change in climate compared to the planet's historical climate is known as climate change (Villalpando, 2024).

Global climate records from the 20th and 21st centuries report significant changes in the global climate system, especially in relation to the average global temperature. During the period 1880-2012, an average increase of 0.85 °C was observed (Sepúlveda, 2020).

The rate and duration of this warming has been greater than at any time in the last millennium; the last 30 years are warmer than any period since 1850. The latest global scientific assessment of climate change, presented in 2013, gives a 95 per cent certainty. It is certain that human activity is the main cause of the observed warming since the mid-20th century (.Martinez and Patiño, 2012).

The report also notes the warming of the atmosphere and oceans, the shrinking of glaciers and snow and ice caps, rising sea levels and increasing concentrations of greenhouse gases (GHGs).

This chapter deals with the components of the global climate system. Taking into account the different greenhouse gas emission scenarios, it describes the factors related to climate change as a consequence of human activities, observed changes and future projections, according to the different emission scenarios.

Climate change

Over the last two decades, climate change has been a topic of daily conversation and concern for many economic and social sectors, but it is also a topic that is not considered important, not only for many people, but also for some economic sectors. Climate change affects all economic and social sectors, certainly including agriculture and food production. Possibly one of the reasons why it is not given due importance is because it is not really understood what it is and how it affects each economic sector. This article presents the causes of climate change, how it impacts on agriculture and some measures to reduce its impact on agricultural production (Villalpando. 2024).

What is climate change?

Climate change has its origins in the increased emission of greenhouse gases at the beginning of the industrial era. This situation has become more pronounced since the 1970s. The greenhouse gases causing global warming are mainly CO_2, methane and nitrogen oxides. The agricultural sector contributes about 1/3 of the total greenhouse gas emissions, through the use of nitrogen fertilisers, deforestation, livestock and rice cultivation by flooding. FAO (2010).

While agriculture is also a sector that helps sequester CO_2 by planting annual and perennial crops, it could further increase its contribution through reforestation and reduction of nitrogen fertilisers. Global warming causes, in the long term, a gradual increase in temperature in the atmosphere near the earth's surface, and in the short term, an alteration of weather patterns, mainly temperature and precipitation. This climatic disorder is manifested by the occurrence of extreme rainfall and temperature events, of magnitude and at times of the year that we are not used to observing, FAO (2007).

In the short term, the effect of global warming on Africa's climate manifested itself with an increase in temperature, starting around the 1990s, and this trend continues to increase to date.

In terms of rainfall, the changes observed are:

1) increased variability in rainfall totals between years in several regions of the country, leading to periods of drought, and
2) greater variability in seasonal and annual distribution.

This situation affects and will continue to affect agricultural production under both storm and irrigated conditions, as gas emissions continue to increase, despite the international commitments on the subject agreed by most countries. In short, climate change is the result of an alteration of the climate, which no longer follows the traditional patterns we were used to observing and is instead much more variable.

2.5. Impact of climate change on agricultural production (Miyamoto, 2017; IPCC, 2014).

Climate change affects agricultural production in different ways. The impact is often negative on:

I) more variable rainfall in space and time, leading to water shortages for rainfed and irrigated agriculture as well;

II) occurrence of high temperatures that cause heat stress to crops and increase water demand. In other cases, however, climate change can be beneficial, e.g. increased CO_2 increases the production of crop species such as fruit trees, wheat, beans and others, as long as the crops have sufficient soil moisture and good nutritional conditions.

In addition, the increase in average temperature in some regions has allowed the agricultural frontier for some crops, such as lemons and avocados, to expand into temperate regions that are now cooler.

Several examples of the impact of climate change on agricultural production are listed below.

- ✓ More variable storms due to climate change impact storm crops because of their irregular onset, duration and the amount of rainfall they bring.
- ✓ Irregular rainfall also impacts the use of agricultural inputs such as fertilisers, crop varieties and the use of chemicals to support crops, among others.
- ✓ Increased variability and scarcity of rainfall during the crop cycle causes water stress in plants, shortens the growing season and reduces yields, especially if it occurs during flowering and full flowering of food crops.
- ✓ The scarcity and irregularity of rainfall is more pronounced on soils with low moisture retention and low levels of organic matter.
- ✓ Increased rainfall variability impacts the water supply of dams and their water availability for irrigated agriculture.
- ✓ Increased temperature shortens the vegetative cycle of crops (days to flowering, maturity) and eventually reduces yields.
- ✓ Increased maximum temperature causes heat stress at different levels, e.g. days with a maximum temperature above 30 °C cause heat stress to crops such as avocado, wheat and others. Hot days also increase the water demand of crops.
- ✓ The increase in minimum temperature reduces the availability of chilling hours for fruit crops such as apple, peach, walnut and others, as well as for winter wheat production, especially in the region of Sonora and Baixio. Increased minimum (night) temperatures also negatively impact other crops grown in tropical regions, such as mango and maize.
- ✓ The occurrence of rainfall associated with higher temperatures at inopportune times, due to climate change, is often associated with the emergence of diseases and pests.

Climate change adaptation measures to reduce the impact of climate change on agricultural production

1. Agricultural production in the face of climate change requires adaptation measures to reduce vulnerability in food production.
2. What can farmers and policymakers do in the face of such a paradigm?
3. First of all, it must be said, agricultural production can and should no longer be carried out according to the recommendations of a "technological package" that is not linked to climatic information. Sowing dates, crop varieties, protection practices against pests, diseases and ailments, the amount of irrigation water and fertilisation levels will, to a greater or lesser extent, be affected by the climatic conditions experienced during the growing cycle. Therefore, farming should no longer be practised by repeating the same "technology package" every year.
4. Growers should expect regular information on expected weather conditions for each growing season. This information includes seasonal and monthly weather forecasts, the occurrence of El Niño or A Menina weather events and 7-day weather forecasts during the growing season.
5. Promote soil moisture conservation practices and the incorporation of organic matter into the soil as medium and long-term measures to mitigate the impact of rainfall variability.
6. Use crop varieties tolerant to water or heat stress according to expected climatic conditions, as an adaptation measure to reduce the impact on production.

Over the past few decades, Africa has seen an increase in climate variability, rising temperatures and a significant increase in the number or intensity of droughts, which have not only doubled in frequency of occurrence, but have also seen a considerable increase in the number of extreme eventsHernandez (2014) and Promis and Ibarra (2010).

The observed trends in Africa's climate had an appreciable negative impact on a wide range of socio-economic activities, most notably on the agricultural sector. The extreme events associated with ENSO (O Menino - Southern Oscillation 97/98), had a similar effect on the agricultural sector and were one of the causes of a reduction in the production of groceries and other items of great economic interest for the country (METRO Enrique, 2015) and Oñate (2012).

These notable adverse events were occurring more frequently during the country's main agricultural period (cold season). However, there are crops and livestock that benefit depending on the magnitude of these extreme events and the stage of development they are in. There is no doubt that locally intense rainfall damages most established crops at a given time, however, water from runoff collected in a catchment can ensure that new crops are harvested (UNEP, 2008).

Concerns related to the above-mentioned aspects guided a whole series of studies aimed at understanding the causal agents of the disasters that were occurring in agriculture and, subsequently, the establishment of monitoring systems that, while not acting in the face of extreme events, provided the growing culture among users of meteorological information.

The issuing of early warnings on extreme events succeeded in incorporating meteorological knowledge into the tactics and operational measures that agricultural decision-makers must routinely take to mitigate and minimise the adverse effects of weather and climate. SegundoOñate (2012), The early adaptation strategy is reaffirmed as one of the best options for mitigating climate change.

SegundoSuárez (2009), in concluding their studies on Climate Variability, Impacts and Adaptation, noted that "it is of paramount importance for society to deepen awareness of the different elements of climate variability in Africa, its extremes and the magnitude of its impacts. Only then will there be a better chance of predicting its impacts and, consequently, a better chance of establishing development plans that take these factors into account.

Drought is the disaster that affects the largest number of people worldwide. It can cause mass migration to cities, contribute to food insecurity and affect the supply of food to the population. It also affects services in different sectors, mainly health and education, and promotes potentially dangerous conditions for vegetation fires Hassan (2010).

The integration of theoretical concepts on the evolution of agricultural drought and potential fire hazard conditions in vegetation, extended through the application of Geographic Information Systems to all regions of the country, constitutes a valuable tool that allowed the assessment of the climatic risk caused by the presence of these extreme weather events produced by climatic variations, Gutiérrez and Pons (2006).

2.6. Angola's climate (Development Workshop, 2011).

Angola is a country in intertropical Africa, with a climate marked by alternating seasons, with a dry season from May to September and a rainy season from October to April. This peculiar geographical situation is due to the fact that Angola is located in the intertropical and subtropical zone of the southern hemisphere and close to the sea. The cold and dry Benguela Current generates rare rainfall, which extends from the coast to the central plateau. Temperatures in Angola are highest during the rainy season, peaking in March.

The dry season represents the cold season, commonly known as cacimbo, and peaks in June and July. The cold is less intense on the coast and increases in intensity as one moves towards higher altitudes. On the central plateau, the temperature drops below 10 degrees Celsius and can reach 5 degrees Celsius. The average annual temperature is 27 degrees Celsius and the minimum is 17 degrees Celsius. This climatic diversity corresponds to the tourist potential represented by a natural heritage rich in diverse flora and fauna, which allows the practice of all kinds of activities, such as leisure and adventure.

Climate and average weather all year round in Ondjiva, Angola(MTTICS, 2019)

In Ondjiva, the rainy season is mostly cloudy, the dry season is mostly clear and it is very hot all year round. Throughout the year, the temperature generally varies from 10 °C to 37 °C and rarely falls below 7 °C or rises above 39 °C.

In support of the beach/pool score, the best times of the year to visit Ondjiva for warm weather activities are from 21 April to 29 April and from late August to mid-October.

Spring weather in Ondjiva Angola.

Source:https://pt.weatherspark.com/s/78292/0/Condi%C3%A7%C3%B5es-meteorol%C3%B3gicas-m%C3%A9dias-na-primavera-em-Ondjiva-Angola.12- 04-2024.

Daily maximum temperatures hover around 35 °C, rarely dropping below 30 °C or exceeding 39 °C. The highest average daily maximum temperature is 37 °C on 11 October (Figure 1).

Daily minimum temperatures rise by 6°C, from 15°C to 21°C, and rarely fall below 12°C or rise above 23°C.

For reference, on 10 October, the hottest day of the year, temperatures in Ondjiva generally range between 19°C and 37°C, while on 5 July, the coldest day of the year, they range between 10°C and 28°C. .

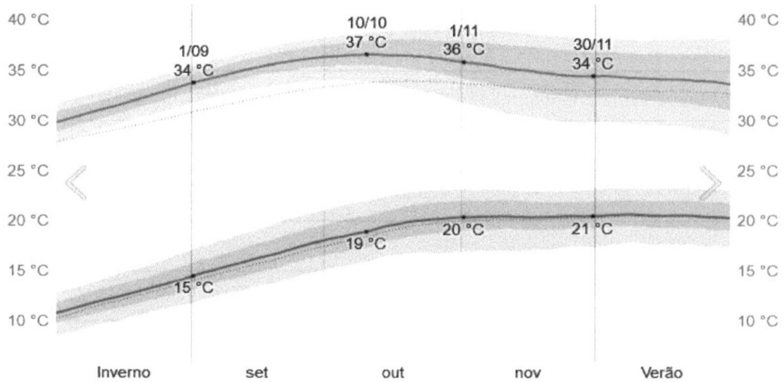

Mean maximum (red line) and minimum (blue line) temperatures, with ranges from the 25th to the 75th percentile and from the 10th to the 90th percentile. The thin dotted lines are the corresponding perceived mean temperatures.

Figure 1: Average maximum and minimum spring temperature

The following figure (figure 2) shows a compact characterisation of hourly mean temperatures during spring. The horizontal axis indicates the day and the vertical axis indicates the hour. The colour is the average temperature for that hour of that day.

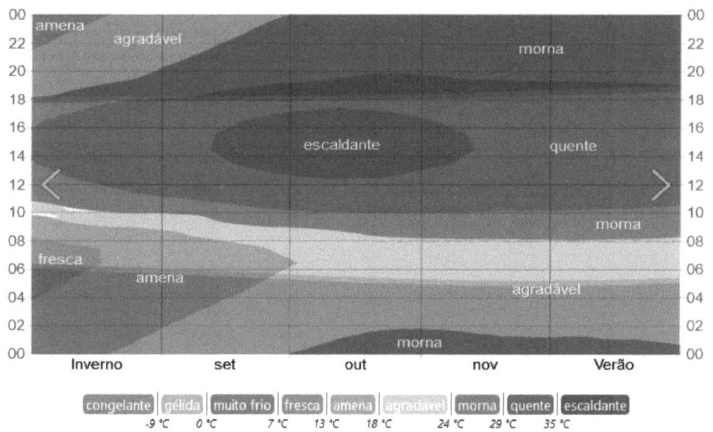

The hourly mean temperature, coded in colour bands. Civil twilight and night are indicated by shaded areas.

Figure 2: Average hourly temperature in spring

Clouds

In spring in Ondjiva, cloudiness (Figure 3) increases dramatically. The percentage of time that the sky is cloudy or mostly cloudy increases from 11% to 58%.

The clearest day in spring is 5 September, with clear, mostly clear or partly cloudy skies 89% of the time. For reference, on 28 January, the cloudiest day of the year, the probability of overcast or mostly cloudy skies is 73%, while on 12 June, the least cloudy day of the year, the probability of clear, almost clear or partly cloudy skies. increases to 93%.

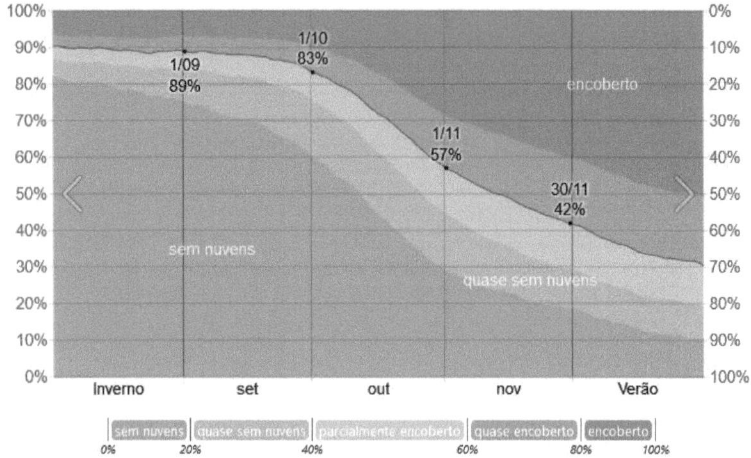

The hourly mean temperature, coded in colour bands. Civil twilight and night are indicated by shaded areas.

Figure 3: Cloud cover categories in spring

Precipitation

A day with precipitation is considered to be a day when the minimum liquid precipitation or liquid equivalent is 1 millimetre. In Ondjiva, the probability of a wet day during spring increases very rapidly, starting at 0 % and ending at 24 %. As a reference, the highest yearly probability of having a wet day is 43 % on 26 January, and the lowest probability is 0 % on 22 July. (figure 4)

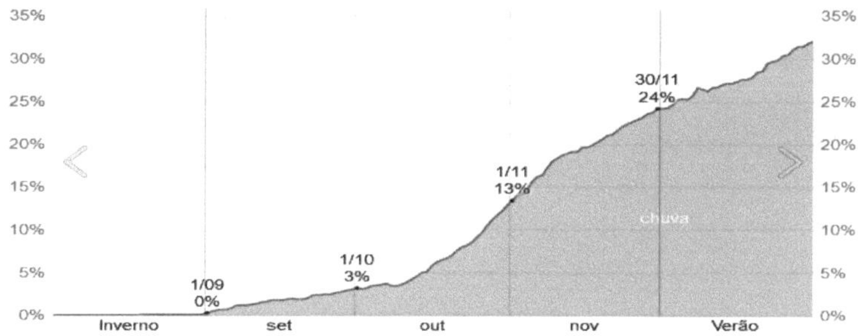

Percentage of days on which several types of precipitation are observed, except for insignificant amounts: rain only, snow only and mixed (rain and snow on the same day).

Figure 4: Precipitation probability in spring

Rain

To demonstrate within-season variation and not just monthly totals, we show the cumulative rainfall over a continuous 31-day period around each day (Figure 5). The 31-day moving average rainfall during the spring at Ondjiva increases rapidly, starting the season with 2 millimetres, when it rarely exceeds 5 millimetres, and ending the season with 15 millimetres, when it rarely exceeds 35 millimetres, when it rarely exceeds 3 millimetres. inches, or less than 6 millimetres.

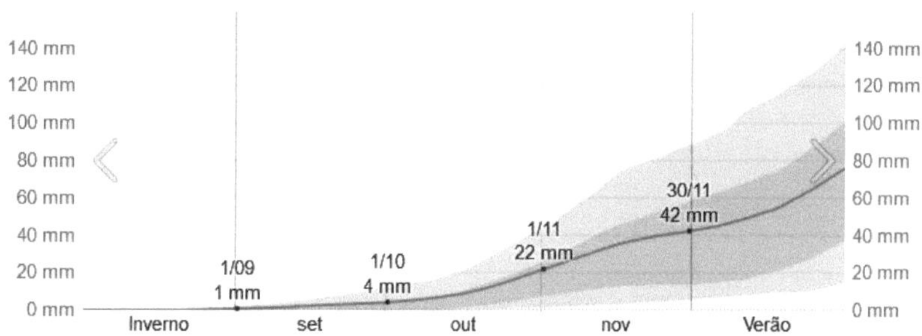

Average precipitation (solid line) accumulated during the 31-day continuous period surrounding the day in question, with ranges from the 25th to the 75th percentile and from the 10th to the 90th percentile. The thin dotted line is the corresponding average snowfall.

Figure 5: Average monthly precipitation in the spring

Sol

During the course of spring in Ondjiva, the length of the day increases rapidly. From the beginning to the end of the season, day length increases by 1 hour and 17 minutes, resulting in an average daily increase of 51 seconds and a weekly increase of 5 minutes and 59 seconds. The shortest day in spring is 1 September, with 11 hours and 47 minutes of sunlight. The longest day is 30 November, with 13 hours and 4 minutes of sunlight (Figure 6).

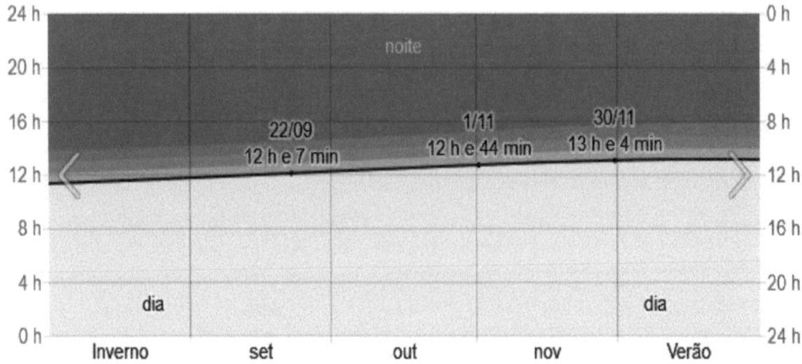

Number of hours the sun is visible (black line). From bottom (more yellow) to top (more grey), the coloured bands indicate: total sunlight, twilight (civil, nautical and astronomical) and total night.

Figure 6: Daylight hours and twilight in the spring

The latest sunrise in spring at Ondjiva is at 6:03 on 1 September and the earliest sunrise is 50 minutes earlier at 5:13 on 24 November. The earliest sunset is at 5:50 p.m. on September 1 and the latest sunset is 27 minutes later at 6:17 p.m. on November 30. Daylight saving time is not observed in Ondjiva during 2024.

For reference, on 21 December, the longest day of the year, the sun rises at 05:20 and sets 13 hours and 9 minutes later at 18:29, while on 20 June, the shortest day of the year, the sun rises at 6:25 am and sets 11 hours and 7 minutes later at 5:32 pm.

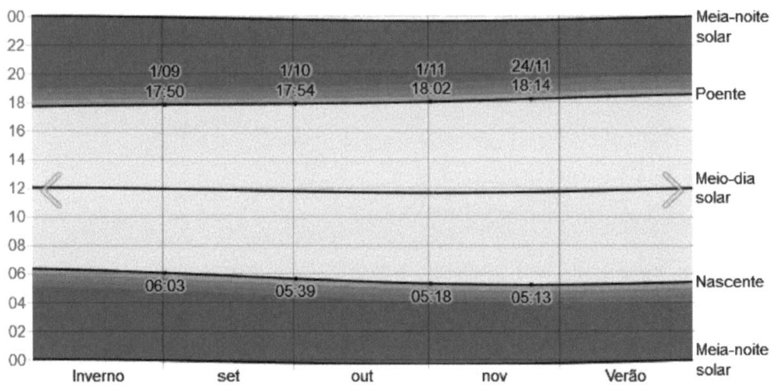

The solar day in spring. From bottom to top, the black lines are the previous solar midnight, sunrise, solar noon, sunset and the following solar midnight. Day, twilight (civil, nautical and astronomical) and night are indicated by colour bands ranging from yellow to grey.

Figure 7: Sunrise and sunset with twilight in the spring

The following figure (figure 8) shows a compact representation of the sun's elevation (the angle of the sun above the horizon) and azimuth (the compass reading) for each hour of each day in the reporting period. The horizontal axis indicates the day of the year and the vertical axis indicates the time of day. For each day and time of day, the background colour indicates the azimuth of the sun at that time. The isolines are contours of constant solar elevation.

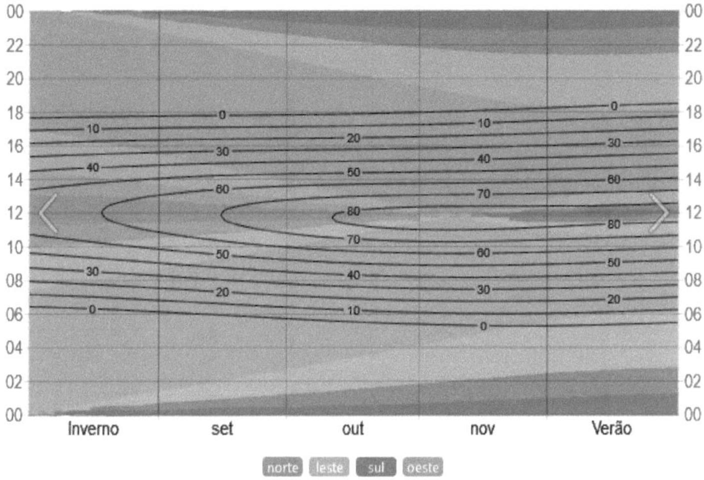

Solar elevation and azimuth in spring 2024. The black lines are lines of constant solar elevation, i.e. the angle of the sun above the horizon, in degrees. The coloured backgrounds indicate the azimuth (compass reading) of the sun. The fainter areas at the cardinal point boundaries indicate the implied intermediate directions (northeast, southeast, southwest and northwest).

Figure 8: Solar elevation and azimuth in spring.

Moon

The following figure (figure 9) is a compact representation of important lunar data for spring 2024. The horizontal axis indicates the day and the vertical axis indicates the time of day. The coloured areas indicate when the Moon is above the horizon. The grey vertical bars (New Moons) and the blue bars (Full Moons) indicate the main phases of the Moon. The information associated with each bar indicates the day and time at which the phase is reached and the related hourly information indicates the times of the phases. The Moon rise and set at the closest time interval when the Moon is above the horizon.

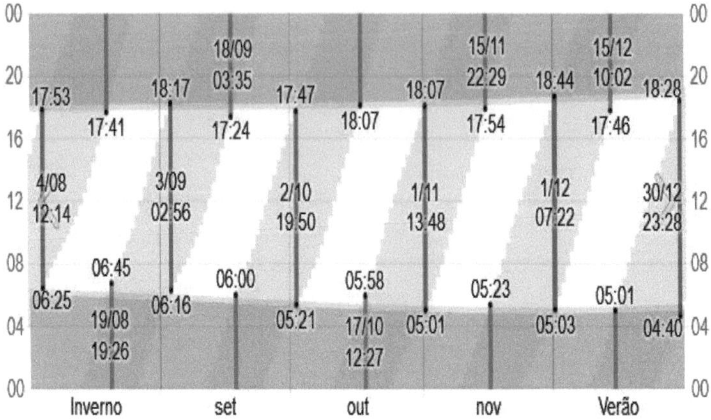

The time the Moon remains above the horizon (light blue area), as indicated by New Moons (dark grey lines) and Full Moons (blue lines). Civil twilight and night are indicated by shaded areas.

Figure 9: Moonrise, moonset and moon phases in spring

Humidity

We base the level of moisture comfort on the dew point, as it determines whether perspiration will evaporate from the skin and consequently cool the body. Lower dew points result in a drier feeling. Higher dew points result in a more humid feeling. Unlike temperature, which generally varies significantly from day to night, the dew point tends to change more slowly. So, while the temperature may drop at night, a muggy day is usually followed by a muggy night.

The probability of a given day being sultry in Ondjiva increases gradually during spring, rising from 0 % to 3 % over the course of the season. For reference, on 6 March, the sultry day of the year, conditions are sultry 25% of the time. On 21 May, the least sultry day of the year, conditions are sultry 0% of the time (Figure 10).

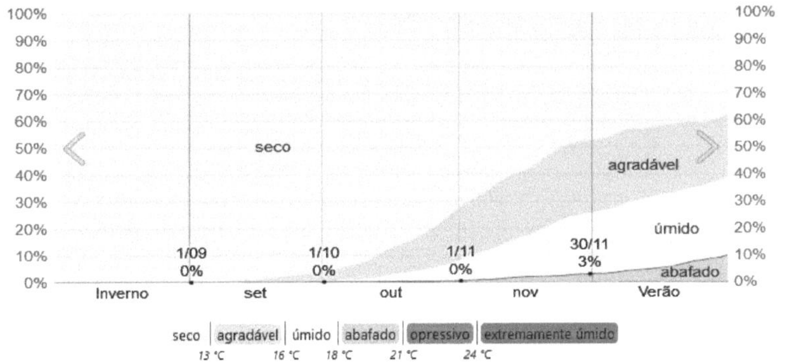

The percentage of time spent at various humidity comfort levels, categorised by dew point.

Figure 10: Moisture comfort levels in the spring

winds

This section analyses the average hourly wind vector (speed and direction) over a wide area 10 metres above the ground. Wind sensation at a given location is highly dependent on local topography and other factors. Wind speed and direction at an instant vary much more than hourly averages (Figure 11).

The average hourly wind speed in Ondjiva decreases rapidly during spring, with a decrease from 15.1 kilometres per hour to 11.4 kilometres per hour over the course of the season. For reference, on 15 July, the day with the strongest winds of the year, the average wind speed for the day is 15.3 kilometres per hour, while on 7 March, the day with the least winds of the year, the average wind speed is 15.3 kilometres per hour. The daytime wind speed is 10.6 kilometres per hour. The maximum daily average spring wind speed is 15.3 kilometres per hour on 7 September.

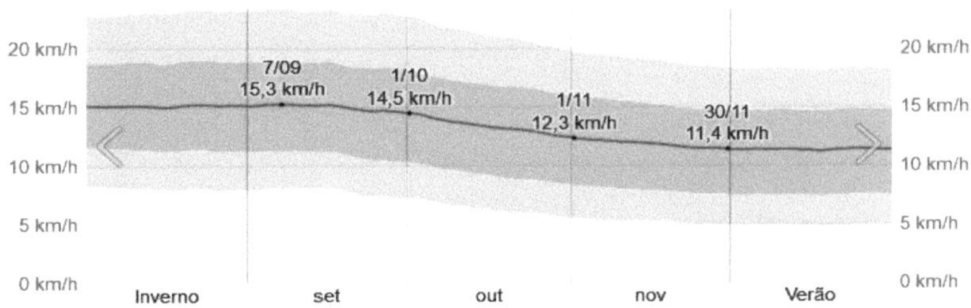

Hourly mean wind speed (dark grey line), with ranges from the 25th to the 75th percentile and from the 10th to the 90th percentile.

Figure 11: Average wind speed in spring

Figure 12 shows that the average hourly wind direction at Ondjiva in spring is predominantly from the east, with the highest proportion of 64 % on 1 September.

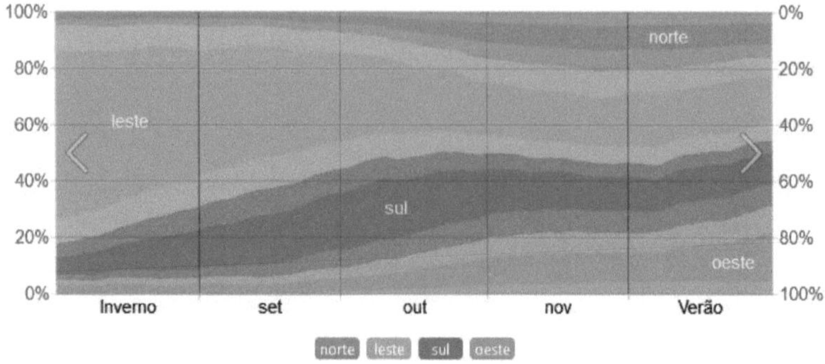

The percentage of hours when the wind has an average wind direction from each of the four cardinal directions, except for hours when the average wind speed is less than 1 mile per hour. The greyer areas at the intersections indicate the percentage of hours spent in the implied intermediate directions (northeast, southeast, southwest and northwest).

Figure 12: Wind direction in spring

Growing season

Definitions of the growing season vary around the world. For the purposes of this report, our definition is the longest continuous period of the year in which temperatures are free of frost (≥ 0°C) (the calendar year in the northern hemisphere and 1 July to 30 June in the southern hemisphere). Temperatures in Ondjiva are warm enough throughout the year that it makes sense to analyse the growing season from that perspective. However, we have included the following graph as an illustration of the temperature distribution throughout the year (figure 13).

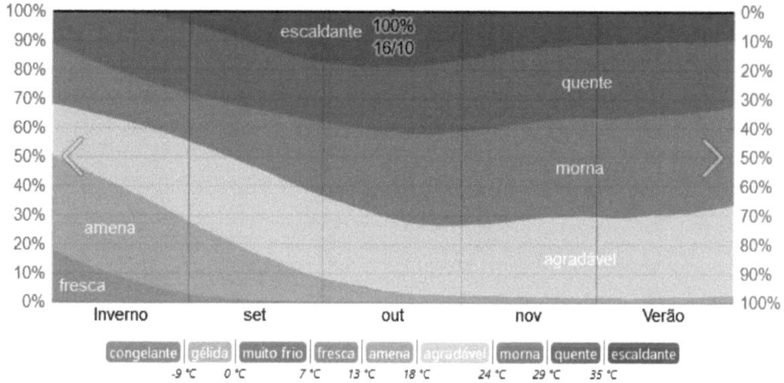

The percentage of time spent in the different temperature ranges. The black line is the percentage probability that a given day belongs to the growing season.

Figure 13: Time spent in different temperature bands and spring growing season

Degree days are a measure of annual heat accumulation used to predict the development of animals and plants, defined as the integral of the heat above the base temperature, discarding any excess above a maximum temperature. In this report we use a base of 10°C and a maximum of 30°C. The average cumulative growing degree days at Ondjiva increase very rapidly during the spring, rising by 1,404°C, from 616°C to 2,020°C over the course of the season (figure 14).

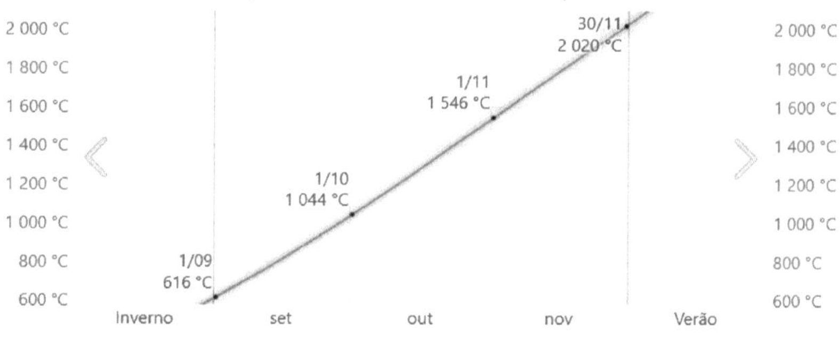

Average cumulative degree days during the spring, with percentile ranges from 25 to 75 and from 10 to 90.

Figure 14: Growing degree days in springWinter

Solar energy

This section analyses the total daily incident shortwave solar energy reaching the ground surface over a wide area, taking into account seasonal variations in day length, elevation of the sun above the horizon and absorption by clouds and other atmospheric elements. Shortwave radiation includes visible light and ultraviolet radiation . The average daily incident shortwave solar energy at Ondjiva increases gradually in the spring, rising by 0.6

kWh, from 6.5 kWh to 7.2 kWh over the course of the month. The highest daily average incident shortwave solar energy in spring is 7.4 kWh on 11 October (Figure 15).

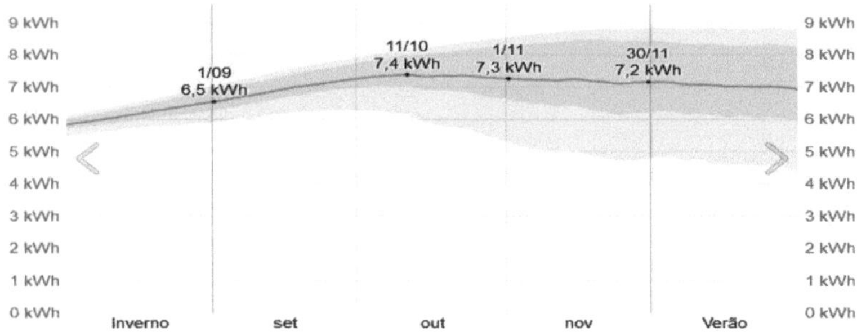

Average shortwave solar energy reaching the ground (orange line), per square metre, with ranges from the 25th to the 75th percentile and from the 10th to the 90th percentile.

Figure 15: Daily average incident shortwave solar energy in the spring

Topography

For the purpose of this report, the geographical coordinates of Ondjiva are latitude: -17.067°, longitude 15.733° and elevation: 3500 feet. The topography within a 3-kilometre radius of Ondjiva is basically flat, with a maximum elevation difference of 11 metres and an average altitude above sea level of 1,114 metres. Within a perimeter of 16 kilometres, the surface is also basically flat (28 metres). Within a perimeter of 80 kilometres, the area is basically flat (85 metres). The area within a 2-mile radius of Ondjiva is covered by grassland (92%); within a 16-kilometre perimeter, by pasture (90%). Finally, within the 80-kilometre perimeter, by pasture (81%).

About the term information system

The purpose of information and its importance as an indispensable resource for society has been addressed by multiple authors in different circumstances and contexts, as it has become an essential factor for human action in modern society (Harrell, 2001).

The specialised literature presents a diversity of definitions of this term due to the different approaches to human activity, which resulted in a lack of consensus among authors to find a single, formally recognised or generalising definition .

According to Rosales (2001), this diversity of definitions may be due to factors related to the human being, when developing his ideas, theories and concepts, being under the influence of his personal experience in the field of knowledge where he is involved and, therefore, presents different points of view; the context in which the studies are carried out (time, space and concrete conditions), which impose different perceptions among subjects; and differences in the intentionality or implicit attitude of people whose purposes are specific.

According to Pernía and Fornés (2009), an information system is an integrated set of processes, mainly formal, developed in a user-computer environment that, operating on a set of structured data (database) of an organisation, selectively collect, process and distribute the information necessary for the normal functioning of the organisation and its management activities.

Information is an important agent in the modification of existing behaviour in the organisation; its correct management is a fundamental tool for decision-making, staff training, product evaluation, error determination and process control, hence it is considered a vital intangible resource for the development of the organisation (Villalpando, 2024).

Develop and regularly update a food insecurity and vulnerability information system that indicates the areas and populations at the local level suffering or at risk of hunger and malnutrition and the elements contributing to food insecurity, making maximum use of existing data and other information systems, enabled the development of the VAM, Analysis and Mapping of Vulnerability to Food Insecurity in Angola (Montenegro and Lujan, 2018).

The development potential of this territory is based on the results offered by the diagnostic indicators seen from the perspective of resilience, i.e. the capacity of the agricultural ecosystem to recover, taking into account the degree of exploitation to which it has been, is or will be subjected, and is closely related to the vulnerability of the agricultural ecosystem, which can be assessed on a scale of High, Medium and Low (UNDP,2006).

Vulnerability: the probability of a sharp decline in food access or consumption levels below certain critical values. In the terms and purposes with which WFP fulfils its mission and orients its actions towards food insecurity vulnerability assessments, vulnerability is defined as an additional measure, for a given population or region, of the risk of exposure to food insecurity and of the population's capacity to cope with its consequences, the overall problem to be solved is underpinned by the assessment and is expressed as:

VULNERABILITY = RISK + RESPONSE CAPACITY

Risk. These are negative or harmful processes that occur as a consequence of natural phenomena themselves, or of the reaction of the environment to the inadequate handling that man can often do with it.

Risk Factors: Choice of those events that constitute the situations of greatest impact in the region and that are determinant in the agro-economic situation (Montenegro and Luján, 2018).

In this regard, risks arising from extreme weather events are identified, namely:

- ✓ Droughts, as phenomena that generally result from extreme imbalances between the evaporation variable and the precipitation variable over time and that have shown more common manifestation rates over longer periods.
- ✓ Desertification processes, associated with the advance of phenomena that degrade soil resources.
- ✓ Characteristics of the relief, or the anthropic transformation of the environment that produces losses in the natural drainage capacity of the territory or in the absorption capacity of the soil, with the occurrence of meteorological events that generate heavy rainfall.

The selection of these risk factors was based on the analysis of their probability of occurrence, intensity, spatial extent and other particular characteristics of the phenomenon, according to the available historical series and the specific information recorded in this regard (Villalpando, 2024).

The phenomenon of drought, in the first instance, is a climatic phenomenon caused by a reduction in precipitation that manifests itself slowly, affecting people, economic activities and the environment in general, it is an event that occurs when the availability of freshwater, from rain, rivers, ponds, underground shells (all sources), is continuously below the usual values.

In the International Meteorological vocabulary (Montenegro and Luján, 2018), drought is defined as "a period of abnormally dry weather conditions, sufficiently prolonged that the lack of precipitation causes a severe hydrological imbalance. Flood risk. Floods, which are generally caused by torrential rains accompanying certain weather events, are one of the main causes of significant damage to agriculture and food production in some areas.

Other associated problems, such as soil modifications caused by inappropriate agricultural practices, logging, fires, urbanisation and other inappropriate environmental interventions or combinations thereof, contribute to increased flood risk. Response capacity.

It should be pointed out that this index depends exclusively on climatic factors and does not take into account the hydrophysical characteristics of soils and vegetation. With all this information, a ten-yearly Agro-meteorological Monitoring Bulletin would be produced, which would be a very important tool for decision-making in the agricultural sector. Elaboration of a ten-yearly agro-climatic monitoring bulletin that gathers the behaviour of the current climatic variables and provides perspectives for the next decade, as well as the agro-meteorological conditions for the different crops.

III. MATERIALS AND METHODS

3.1. Location

Cuanhama is a municipality in the province of Cunene in Angola, which has its headquarters in the city of Ondijiva, making it the capital of the province. It covers an area of 20,255 km^2 and has about 211,000 inhabitants. It is bordered to the north by the municipality of Cuvelai, to the east by the municipality of Menongue, to the south by the municipality of Namacunde and to the west by the municipality of Ombadja. The municipality is made up of the headquarters commune, which corresponds to the city of Ondijiva, and the integration of the communes of Môngua, Evale, Nehone Cafima and Simporo.

3.2. Characterisation of the agricultural development of the municipality (AIPAEX, 2024)

the climate

Semi-desert, dry tropical; megathermal, with irregular rainfall not exceeding 600 mm per year. The average annual temperature is 23 degrees Celsius, with large daily variations in temperature. The highest concentration of rainfall occurs between December and April, with large irregularities in its distribution.

Agricultural activity

Agriculture is rain-fed, based on massango and massambala crops. There are currently 51,650 family farms in the province with a total area under cultivation of 77,475 ha. Of which 43% is cultivated annually. Cattle rearing is well known as the most important productive activity in Cunene, estimated at over 1,000,000 head of cattle. Most of the herds are owned by traditional breeders and the herd is maintained extensively using natural pastures which determine the availability of water in the first place and the potential for grazing stock. Periodic limitations of one of these factors lead to movements of cattle herds to transhumance areas that vary in space and time.

Artisanal fishing plays an important role in supplying fish to rural populations and contributes to improving the diet of the communities. This activity is mainly practised in the Cunene River with low catch rates due to the weak support of fishing gear (lines, hooks, buoys, weights, nets, etc.).

This research is part of the project "Comprehensive Participatory Assessment of Family Farming and Livestock in the Province of Cunene, Angola" (DIPAF), which is being carried out in four municipalities (Namacunde, Cuvelai, Ombanja and Cuanhama) in the Province of Cunene, executed and coordinated by the Polytechnic Institute of Ondjiva.

The purpose is to expose the methodological procedure that allows the improvement of the Information System of climatological variables in the municipality of Cuanhama so that it constitutes a tool for decision making by social decision makers. The following methodological steps are suggested to carry out the research:

1. Selection of the study area.
2. Information gathering.
3. Information processing and analysis.
4. Production of the ten-yearly bulletin containing an information system of climatological variables.

3.3. Selection of the study area.

Cuanhama's selection criteria are related to its condition as a rural municipality, the predominance of agricultural activities, the impact of adverse weather events that make it a fragile ecosystem with high vulnerability to drought, desertification and flooding.

On the other hand, the interest of the governmental authorities of the territory in solving the problem of regional development with a sustainable approach based on agricultural activity. These criteria correspond to the objectives of the DIPAF project.

3.4. Information gathering

To collect information, social decision-makers were interviewed; secondary information had to be sought through the review of documents, soil data and studies carried out by other organisations. The following meteorological variables and their modalities were defined:

a) Variable temperature: This is the value recorded at the time of the reading, it is an indicative index of the heating or cooling of the air resulting from the heat exchange between the atmosphere and the earth, it indicates in numerical values the level of internal energy found in a place at that moment, in addition, it is in equilibrium between the system (plants, animals, etc.) and the environment (air) (Villalpando, 2024).

- ✓ *Average maximum temperature:* is obtained from the maximum recorded within the 24-hour hourly records arising from the thermograph band reading. These records are compared with the maximum reading of the mercury thermometer.
- ✓ *Average minimum temperature:* is obtained from the minimum recorded within the 24-hour hourly records arising from the thermograph band reading. These records are compared with the minimum alcohol thermometer reading.

b) Variable relative humidity: humidity is measured by the water content in the atmosphere, measured with a hygrometer. Pernía and Fornés (2009).

(c) Precipitation: any amount of meteoric water collected on the earth's surface (Villalpando, 2024).

3.5. Data processing and analysis.

To process the information on each variable and its modalities, an Excel database was created with information collected on the climatic variables for the ten years analysed and the year of verification, see (Annex 1), which allowed the creation of line graphs with markers in order to show the trends over time and their behaviour during the period analysed.

3.6. Production of the ten-yearly bulletin containing an information system of climatological variables.

This bulletin aims to develop technical information capacities in the exploration of climate activities, in order to contribute to the improvement of agricultural production systems in the municipality of Cuanhama, without deteriorating the environment.

For the treatment of the data obtained, tools were used that helped in the analysis and processing of results, Microsoft Excel was implemented in the values of variables, indicators and percentages and in the elaboration of figures.

IV. OUTCOME AND DISCUSSION

The behaviour of the climate, especially with regard to the occurrence of extreme events, is of vital interest for human social life. Its consequences are reflected in many different fields of the national economy, but most especially in agriculture, which can be considered as a large open-air factory in which all the activities carried out there depend on weather and climate (Montenegro and Luján, 2018).

4.1. Historical monthly average behaviour of climatological variables in the study area from 2013 to 2022

Maximum temperature

Regarding the historical average maximum temperature variable, corresponding to the period 2013 - 2023 (ten years), it can be observed that the months with the highest temperatures corresponded to the months of January, September, October, November and December with values such as: 35.5, 37.9, 38.9, 38.7 and 37.4 °C respectively and that in the year being compared (2023) the maximum temperature values corresponded to the months of September, October and November with values of 34.1, 37.1 and 33.3 °C (figure 16).

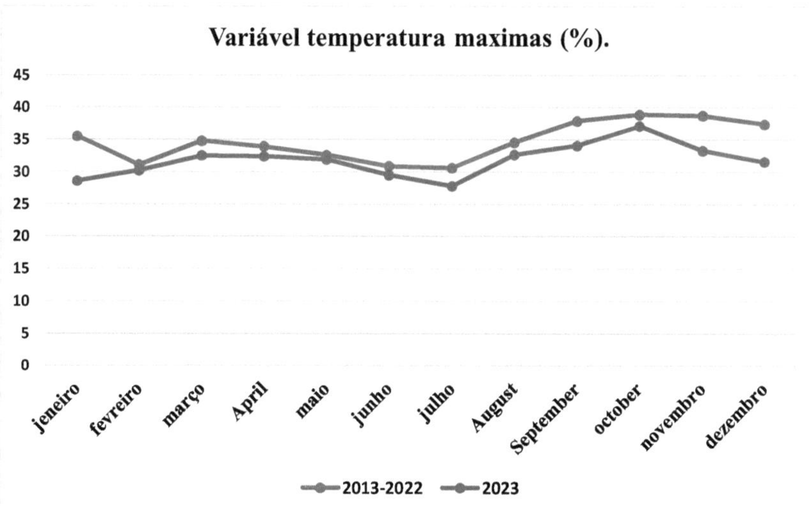

Figure 16: Analysis of the maximum temperature in 2023 in relation to the historical average (2013-2022).

The results shown in the year 2023 in relation to the historical average were lower in all months, with results ranging from 0.9 °C to 6.9 °C, with the historical average temperature being 34.7 °C and in the year analysed 31.8 °C.

Symborg (2024), explains that the incidence of high temperatures is a phenomenon that is here to stay as a consequence of climate change and that negatively affects people, animals and also plants, so that, in crops, the effects of heat are closely linked to The decreases in production are a consequence of the physiological and metabolic stress caused by the crop. Very high temperatures, for example, above 40 °C, can cause crops to go into metabolic shutdown. This metabolic shutdown prevents the transfer of nutrients, which has a very negative effect on the development of the fruit, which will see its growth potential greatly affected. Limited to reach commercial sizes. It is estimated that an increase of 1°C in the average annual temperature can cause a drop in crop productivity of between 4 - 10%.

Furthermore, it is important to explain, according to this author, that "abiotic stress" consists of environmental factors that alter the physiological, morphological and biochemical processes of crops, of these abiotic factors the one that most affects plant stress is temperature, both high and low. Thus, measures should be taken and agricultural practices adopted to reduce the effects of high temperatures, such as, for example, improving water management and using innovative technologies such as biotechnology, with biological solutions developed from microorganisms and biomolecules. Biotechnology, with biological solutions developed from micro-organisms and bio-molecules, helps to ensure crop productivity and profitability in a sustainable and environmentally friendly way, even under adverse conditions.

Where with the use of biostimulants supported by the unique Arbuscular Mycorrhiza Forming Fungus (AMF) Glomus iranicum var tenuihypharum. This fungus establishes a long-lasting symbiotic relationship in which the fungus exchanges water and nutrients with the plant in exchange for sugars derived from photosynthesis, the plant's stomata open for longer, even in extreme conditions. The plant can continue to complete its metabolic cycles and increase the size of its fruit.

4.1.2. Minimum temperature

In the historical average (2013-2022) analysed it is observed that the monthly average in all months was lower than the average shown in the year 2023, value it, but the minimums recorded in the historical average were the months of June, July and August, with values fluctuating between 6.5. 7.6 and 8.7 °C, in relation to the year of comparison (2023) the lowest values also corresponded to the same months (June, July and August) with values of 10.1, 9.9 and 13.5 °C respectively (Figure 17).

The minimum temperature values shown in the comparative year (2023) were higher in all months than those shown in the historical half-year (2013 - 2022), where values fluctuated between 2 and 6.4 °C higher in all months (Figure 17).

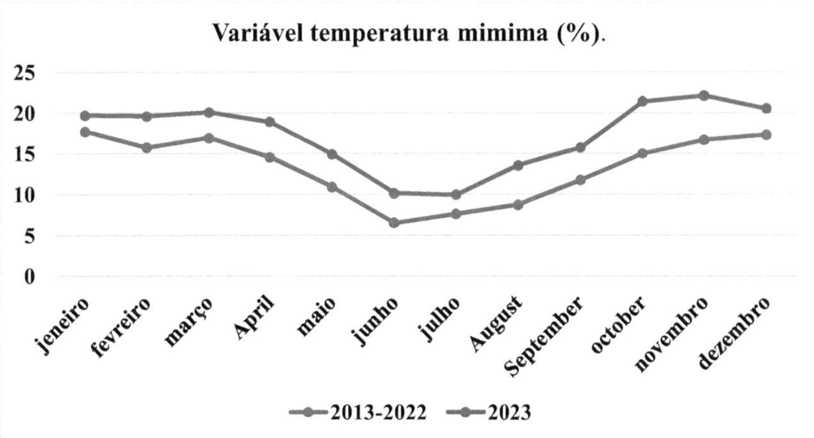

Figure 17: Analysis of the minimum temperature for the year 2023 in relation to the historical average (2013-2022).

This corroborates Vinhas et al. (2024), who stated that in 2023, the Earth's average surface temperature was the warmest on record, according to a NASA analysis. Last year's global temperatures were about 1.2 degrees Celsius (2.1 degrees Fahrenheit) above the average for NASA's reference period (1951 to 1980), reported scientists at NASA's Goddard Institute for Space Research (GISS) in. New York.

This corroborates the increase in minimum temperatures in relation to the historical data analysed, remaining above these, i.e. there is an increase in minimum temperatures, an aspect that influences the development of crops and animals in agriculture.

On the other hand, the FAO. (2024), explains that among the climatic factors that influence cultivation and production is the temperature, which should range between 18 - 25 °C (and the average minimum temperature for the year 2023 was found to be 17.2 °C, which means optimal for crop development), so that the plant can grow properly and produce its fruit with the highest possible quality and yield. Below or above this optimum temperature, the plant cannot develop properly and the crop may not perform its normal biological cycle and may not reach its maximum yield potential.

According to the company Nutricontrol (2020), one of the actions that we must increase today is the production of crops in greenhouses, such as the control of climatic conditions to ensure a better development of the plant, in addition to the temperature (along with other factors such as CO2, lighting, etc.) intervenes in certain functions such as the opening or closing of stomata, which are essential in the vital processes of photosynthesis, transpiration and respiration of plants.

If the temperature inside the greenhouse increases, the amount of water lost through transpiration also increases and the stomata will close as a protective method to prevent excessive water loss. However, this method of support can have a negative effect, as closed stomata restrict the entry of carbon dioxide, which is essential for photosynthesis.

4.1.3. Relative humidity

Relative humidity has little variation, maintaining an annual average of 52.7%, the months with the lowest percentages of relative humidity in the historical average record are August and September (37.0 and 40.3) conditioning the increase in humidity due to the rains that occur in the period, for a similar reason in the winter months the relative humidity is lower (Figure 18).

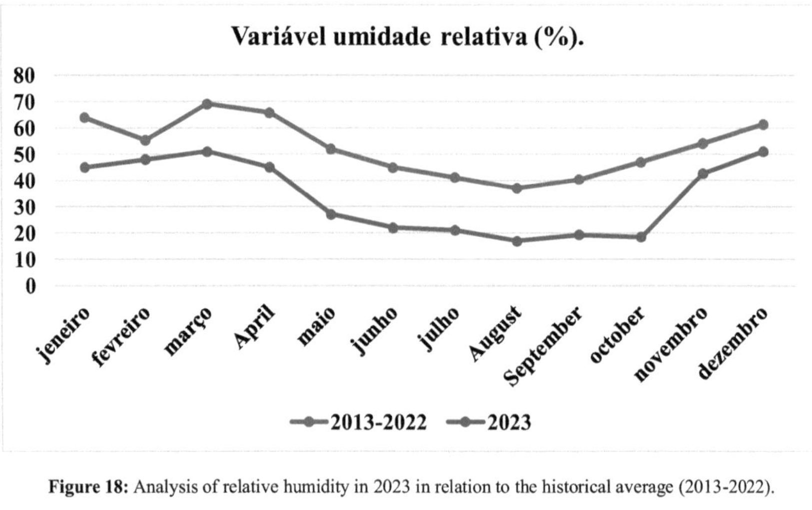

Figure 18: Analysis of relative humidity in 2023 in relation to the historical average (2013-2022).

The relative humidity, although low in the historical average corresponding to the months from June to October (44.9, 41.2, 37.0, 40.3 and 47.0 %) and in the year 2023, the lowest relative humidity values corresponded to the months from May to October (27.0, 22.0, 21.0, 17.0, 19.3 and 18.4%), favourable humidity for the development of the crop.

This corroborates Wikipedia (2013) that relative humidity (RH) is the ratio of the partial pressure of water vapour to the equilibrium vapour pressure of water at a given temperature. Relative humidity depends on the temperature and pressure of the system of interest. The same amount of water vapour produces a higher relative humidity in cold air than in warm air. A related parameter is the cruising point.

4.1.4. Precipitation

Precipitation behaves in an unstable manner (Figure 19): the annual rainfall norm in the historical average territory is 449.5 mm on average, and in the year under study 391.3 mm, which is lower than the historical average. In the period May - October, rainfall was less than 22.9 mm for the historical record and 18.5 mm in the year under study, which is also less than the historical average.

The highest values in both the history (2013-2022) and the year analysed (2023) were found in both cases in the months of January, February, March and December with values fluctuating historically at 78.2, 81.2, 114.2 and 78.4 mm respectively, while in the year of study (2023) the same months behaved as follows: 68.3, 79.5, 97.1 and 68.9 mm, being lower than the historical average in these months.

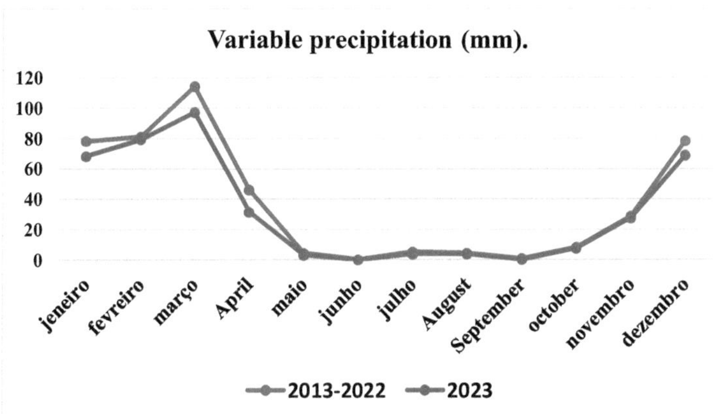

Figure 19: Analysis of precipitation in the year 2023 in relation to the historical average (2013-2022).

As for rainfall, as shown in Figure 19, there is no very high rainfall in each of the months and in both cases it coincides that the months with the lowest rainfall correspond to the months from May to October, with a historical average value of 4.3, 0.1, 0.1, 0.8 and 8.2 mm, with a monthly average of 37.5 mm, 5.2, 4.3, 0.8 and 8.2 mm, with a monthly average of 37.5 mm and in the year 2023 the values were 3.1, 0.0, 3.7, 3.9, 0.2 and 7.6 mm, being less a monthly average of 32.6 mm in the year, a little very unfavourable for agriculture and when in semi-arid areas depends so much on them.

According to Gabo (2023), the amount and distribution of rainfall directly affects the yield and quality of crops and plays a crucial role in water management in agriculture. Plants need water to grow and rainfall is a natural source of this resource. When it rains, soils become wet and plant roots can absorb essential nutrients and water needed for growth. Rain also contributes to photosynthesis, a process vital for plant production. Rainfall not only provides moisture for crops, but also influences water management. In regions where rainfall is seasonal, farmers must carefully plan when to plant and harvest based on rainfall patterns.

Fernández (2016), explains that arid and semi-arid areas are characterised by an average annual precipitation index and a potential evapotranspiration of less than 0.65. This indicates a huge mismatch between the amount of water that can potentially pass into the atmosphere through evapotranspiration and the amount of rainwater that these areas actually receive. These regions have a high atmospheric evaporative demand that depends primarily on solar radiation, air vapour pressure and wind speed; therefore, in most cases, evapotranspiration is limited by the availability of water in the soil relative to atmospheric demand.

The main limiting factor in arid and semi-arid climatic zones is water availability. The quantity and seasonal availability of water is critical for the long-term survival and distribution of plants. Traditionally, the classification of arid, semi-arid and humid zones was based solely on mean annual rainfall (Lloyd, 1986).

According to FAO bulletin 79 (2005), in this case it is very important to apply actions that favour the retention of moisture in the soil, but in these cases of semi-arid areas forces us to take measures to maintain moisture and obtain better plant development, this means that we can keep the crops green and colourful by implementing some simple actions that help to maintain moisture for longer, such as:

1) *Protect crops from wind.* Wind accelerates the dehydration of plants. One way to prevent this is to install breakable curtains, enclosures or other elements that form protective screens.
2) *Make infiltration channels or chimpacas*. This will allow the soil to store water and survive droughts. Gutters are very functional for hanging plants because they reduce water loss by 85%. They also take time for the water to fall, giving it time to be absorbed by the soil, like chimpacas.

3) *Incorporate organic matter.* Periodically supplementing the soil with compost provides organic matter, which increases its porosity.
4) *Maintain what I am used to mulching in summer and winter.* Laying a 5 cm to 10 cm layer of foliage stops moisture loss by evaporation.
5) *Use hydrogels.* Hydrogels, also known as "water retainers", are polymers that have the ability to absorb up to 400 times their weight and produce large quantities of water and other aqueous solutions without dissolving. Their use in orchards, gardens and containers helps to improve the water-holding capacity of soils, favouring plant growth.

Trying to grow native plants is another great tip to keep in mind. These will be able to grow with little water and in the semi-arid soil that characterises our province. In this way we will be making a more rational use of resources, as well as maintaining our environment in its natural harmony.

4.2. Production of the agrometeorological monitoring bulletin every ten years

An information system of climatic variables together with other information reaching the producers is of vital importance to obtain better results in their production.

General objective

Produce a ten-yearly bulletin containing a system of information on climatic variables favourable to producers who can make decisions on their production.

Justification

The Ondjiva locality has areas where different agricultural production systems are found, which are a pillar of both economic activity and local food security. The region is affected by high and often extreme climatic variability, which produces limiting conditions for agricultural production. In addition, the lack of development and adoption of appropriate technologies affects productivity. This is compounded by poor infrastructure and limited access to markets.

The creation of this bulletin is not a mechanism of fixed elements to inform, the fundamental ones are the climatic variables, but in each of them positive aspects were shown so that the producers can improve their productive results, in addition to providing information favourable to agroecological practices. and in favour of the care of the environment, as well as methods, preparation and application of organic and natural fertilizers and insecticides in their productions, among the different aspects that will be informed in this bulletin, which will always be subject to positive comments for its better understanding.

Newsletter

UMN

INSTITUTO POLITÉCNICO DE ONDJIVA

DEPARTMENT OF AGRONOMIC ENGINEERING.

DECENNIAL AGRO-CLIMATIC INFORMATION BULLETIN MUNICIPALITY OF CUANHAMA. PROVINCE OF CUNENE.

<u>3rd DECEMBER JULY/2024</u>

TEN-YEAR SUMMARY

<u>PRECIPITATION.</u>

There were no precipitation records in the decade analysed.

The importance is conditioned by the increase in drought in the territory, aspects that affect the loss of the little water that is stored in the different chimbacas and retained in the soil (the minimum).

In view of these incidents, measures are taken in the different orchards with a view to minimising possible economic losses in the crops due to water stress, measures such as:

el) Apply minimum tillage to crops, avoiding removal of topsoil to prevent moisture increase by evaporation activity.

b) Mulching activities are an important practice in the agronomic management of any crop, especially vegetables, as they are usually small areas established on arid, sandy soils with water scarcity problems. This practice consists of spraying dry grass, foliage or crop residues.

w) Increasing soil organic matter through leguminous cover crops, mulches and compost, rather than synthetic nitrogen, can help absorb rainwater and develop the soil's capacity to retain moisture.

d) Apply survival irrigation, with a view to saving water and being able to reduce water stress in the crop.

y) Protect the soil, avoiding burning the bark of the plant.

F) Maintain strict sanitary measures (disposal of excreta, litter, insect control, etc.).

RELATIVE HUMIDITY:

In the decade it fluctuated between 28.1 - 36.7%, corresponding to a low moisture content.

MAXIMUM TEMPERATURES:

The maximum temperature was between 26.5 and 31.4 °C, averaging around 29.4 °C, a bit high for the favourable development of the crops as the optimum temperature should be between 18 - 25 °C for the plant to grow properly and bear fruit .

The minimum temperature ranges from 7.5 to 17.1 °C, with an average of 12.74 °C.

AGROMETEOROLOGICAL OUTLOOK 1 AUGUST/2024.

1. Similar forecast behaviour in the 1st decade of August/2024.

PRO-AGRICULTURE ACTIONS.

a. Delimit agricultural areas with masts (fruit trees, wood, etc.) to act as windbreaks and thus reduce wind speed between crops.
b. Increasing the use of organic benefits, achieving ecological agriculture.
c. Create and increase the use of natural insecticides from local solutions as preventive methods to prevent pests and diseases.

LIQUID FERTILISERS

Liquid fertiliser is a liquid fertiliser obtained from the degradation of organic matter under anaerobic conditions, it also arises from the liquid preparation made from solid compost or free compost. In addition to its nutritional effect, it is used as a natural pesticide and plant growth stimulator to replenish nutrients during the growing cycle.

One of the easy-to-prepare liquid supplements is manure supplements.

Preparation of dung tea.

When preparing the tea, place 12 kg of any type of manure, place a large stone (to give it weight), tie the bag tightly with string, then place the bag in a container with a capacity of 200 L of water, cover and leave to ferment for two weeks. After this time, the bag is removed, leaving the dung tea inside.

This fertiliser should be diluted with one part manure tea to one part fresh, clean water, and then applied in batches on crops or around fruit trees, as far as the branches extend.

This fertiliser can also be applied via drip line (200 L/ha) every 15 days.

PROTECTING THE ENVIRONMENT IS TRANSCENDING AND GET MORE FOOD.

V. CONCLUSIONS

1. The average maximum temperature of the year under analysis (2023) had a value of 31.58 ºC, being lower than the historical average, which gives an annual average value of 34.7 ºC, the minimum humidity in the year under analysis was 17.2 ºC higher than the historical average (13.3 ºC), the relative humidity of the historical average 52.7% was higher than that of the year under analysis, 34.0%, the rainfall in both cases was low although the historical average (37.5 mm) was higher than that of the year 2023 (32.6 millimetres).
2. An agro-climatic bulletin is produced that provides detailed information on weather conditions in the area, as well as useful elements and information for producers to enable them to make favourable decisions in their production.

VI. RECOMMENDATIONS

1. Establish coordination with the Provincial Meteorological Centre for the systematic updating of the Agro-climatic Monitoring Bulletin every ten years.
2. Extend the information system developed on climatological variables to the Delegation of Agriculture, Government and Provincial Municipality.
3. Continue with the studies of the remaining climatological variables for the creation of the Geographic Information System.

VII. BIBLIOGRAPHY

AIPAEX (2024). Private Investment and Export Promotion Agency. Government of Angola.https://www.aipex.gov.ao/PortalAIPEX/ - !/

World Bank. 2008. Data and statistics. Internet documents. Available at: go.worldbank.org/WVEGH5U9W0

Cambridge, United Kingdom, **Cambridge University Press.**

Cirilo MCM, Candida LL, Yakov FCM (2021). Effect of temperature and precipitation on agriculture in the Coata-Puno watershed, Peru. Journal of Research in Agricultural and Veterinary Sciences. https://doi.org/10.33996/revistaalfa.v5i14.118. May - August 2021. Volume 5, Number 14. ISSN: 2664-0902. pages. 285 - 296

World Commission on Environment and Development (WCED), (1987). Our Common Future. Oxford, United Kingdom, Oxford University Press.

Chavarría. J., Uganda. M., Knowing. A., Munos. J, Bravo. R. and Villalón, A. (2020). Water requirements of cowpea bean (Vigna unguiculata (L.) Walp.). calculated with the crop coefficient using a drainage lysimeter. Science. Agri. 17(3), 111-121. https://doi.org/10.19053/01228420.v17.n3.2020.11746

Clemente Ricse, Jimmy Francisco and DIPAS MEDRANO, Edinson (2016). Effects of Climate Change on the growth rate of Papaya Production in the Mantaro Valley: 2000-2014. P. 73.

Development Workshop (2011). Report on Climate Change Impact in Angola, Vulnerability and Adaptation, Ministry of Environment, Luanda.

Diccionario de lengua española [online] (2020). Available at: https://dle.rae.es/diccionario

Duval, V. and Campo, A. (2017). Microclimatic variations in the interior and exterior of the caldena forest (Prosopis caldenia), Argentina. Cuadernos de Geografía: Revista Colombiana de Geografía, 26(1), 37-49. https://doi.org/10.15446/rcdg.v26n1.42372

FAO. (2007). The State of Food and Agriculture 2007. Rome.

FAO. (2010). Major climate-induced changes. Pp. 9-31.

FAO. (2024). Forest plantations in tropical Africa. https://www.fao.org/4/x5374s/x5374s04.htm#TopOfPage.

Fernando, J. (2020). Salto de Sumba. Universidad del Azuay Facultad de Ciencias de La Administración.

Fernández G, J., (2016). The soil-water resource in arid and semi-arid environments. https://digital.csic.es/bitstream/10261/29768/18/fernandez.pdf

Environmental Information System Fact Sheet. (1972) No. 20. Available at: https://www.dane.gov.co/files/investigaciones/pib/ambientales/Sima/Temperatura_maxima13.pdf.

Gosling, SN and Arnell, NW (2016). Climate change. A global assessment of the impact of climate change on water scarcity, 134(3), 371-385. https://doi.org/10.1007/s10584-013-0853-x

Regional Government of Junín (2010). Regional Climate Change Strategy June. P. 136.

Gutiérrez, Orbeal and Anay, Nerkys. (2019). Behaviour of meteorological variables in psoriasis during 2017.

Gutiérrez, J. and Pons, M. (2006). Numerical modelling of climate change: scientific bases, uncertainties and projections for the Iberian Peninsula. Cuaternario y Geomorfología: Revista de La Sociedad Española de Geomorfología y Asociación Española Para El Estudio Del Cuaternario, 20(3), 15-8.

Guzmán-Castellanos, A.B., Morales-Bojórquez, E. and Balart, EF (2014). Estimating individual growth in elasmobranchs: Inference with multiple models. Hidrobiológica, 24(2), 137-150.

Intergovernmental Panel on Climate Change (IPCC). 2007a. Climate Change 2007: The Physical Science Basis. Contribution of Working Group I to the IPCC Fourth Assessment Report.

Gyoo-Bum, Kim; Han-Heum, Yun; Dae-Ho, K. (2006). Relationship between standardized precipitation index and water table a proposal for the establishment of drought index wells. 2006.pdf. Journal of the Korean Society of Soil and Groundwater Environment, 11, 31-42.

Harrell, F. (2001). Regression modelling strategies: with applications to linear models, logistic regression and survival analysis (Springer-V). New York.

Hassan, RM (2010). Implications of climate change for agricultural sector performance in Africa: policy challenges and research agenda. Journal of African Economies, 19 (suppl_2), ii77-ii105.

Hernández, R. (2014). Metodología de la investigación. 6th Edition. Mexico, 2014. ISBN 9781456223960.

https://pt.weatherspark.com/s/78292/0/Condi%C3%A7%C3%B5es-meteorol%C3%B3gicas-m%C3%A9dias-na-primavera-em-Ondjiva-Angola.12- 04-2024.

https://cultivafuturo.com/beneficios-de-la-lluvia-en-tus-cultivos/ 25-04-2024.

IPCC, IPOCC (2014): Synthesis report on climate change. In: Contributions of Working Group I, II, II to the Fifth Assessment Report of the Intergovernmental Panel on Climate Change. [Sl:sn], 2014.

IPCC, IP on CC (2007a). [Climate Change 2007]: The Physical Science Basics: Summary for Policymakers and Technical Summary and Frequently Asked Questions; As part of the Working Group I contributed to the Fourth Assessment Report of the Intergovernmental Panel on Climate Change. Proceedings of the Alpine Snow. *Workshop, Munich, 5-6 October, Germany HeBIS-Verbundkatalog*, 8, 142. Retrieved from http://bluemarble.nasa.gov.

IPCC (2007b). Climate Change 2007: Impacts, Adaptation and Vulnerability. Contribution of Working Group II to the IPCC Fourth Assessment Report. Cambridge, UK, Cambridge University Press.

IPCC (2007c). Climate Change 2007: Summary Report. Fourth Assessment Report of the Intergovernmental Panel on Climate Change. Geneva, Switzerland.

IDEAM-UNAL (2018). Climate variability and climate change in Colombia, Bogotá, DC ISBN: 978-958-8067-97-1.

Inzunza, Juan (2023). Solar and terrestrial radiation.

Jacto (2023). Agriculture and Management. https://bloglatam.jacto.com/agricultura-depende-pessoas-lluvias/

Jiménez, Rosa María Rodríguez, Capa, Águeda Benito and Lozano, Adelaida Portela (2004). Meteorology and Climatology [online]. 2004. ISBN 8468885355. Available at: https://cab.inta-csic.es/uploads/culturacientifica/adjuntos/20130121115236.pdf

Llacza, Aydee, Hurtado, Femando, Puican, Christian, Barrantes, Kety and Fernández, Alexander (2016). Comportamiento de las variables meteorológicas en el Distrito de Chimbote durante el año 2016 [Thesis]. 2016.

Lloyd, JW, (1986). A review of aridity and groundwater. Hydrological Processes, 1, 63-78.

Maderey Rascón, L.E. and Jiménez Román, A. (2001). Alteration of the hydrological cycle in the lower part of the upper basin of the Lerma River due to the transfer of water to Mexico City. Investigaciones geográficas, 1(45). https://doi.org/10.14350/rig.59143

Martínez, P. and Patiño, C. (2012). Effect of climate change on the availability of water resources. Tecnología y Ciencias Del Água, 3, 5-20.

Manrique, Óscar Brown (2015). Climate change and its precipitation evidence. Hydraulic and Environmental Engineering. 2015. Vol. 36, no. 1, pp. 88-101.

Montenegro, Edgar and Luján Pérez, Marcos (2018). Analysis of the seasonal variation of air pollution and its relationship with climatic variables in the central valley of Cochabamba, Bolivia. Nueva Acta. 2018. Vol. 8, no. 3, pp. 451-466.

Miyamoto, Bruno (2017). "Economic impacts of extreme rainfall on Brazilian agriculture". Thesis submitted to the Unicamp Institute of Economics for a PhD in economic development.

MTICS (2019). Ministry of Telecommunications, Information Technologies and Social Communication. National Institute of Meteorology and Geophysics. Provincial Department of Cunene.

United Nations (2007). Millennium Development Goals. Report 2007. New York, USA.

Núñez, E., Steyerberg, E. W. and Núñez, J. (2011). Strategies for the development of statistical regression models. Revista Española de Cardiología, 64(6), 501-507. https://doi.org/10.1016/j.recesp.2011.01.019

Nutricontrol (2020). The importance of temperature for winter crops. Company dedicated to Research and Technological Development.

Oñate Valdivieso, FR (2012). Project: "Generation of Geoinformation for Land Management at National Level Scale 1: 25,000". In Memoria Técnica. https://doi.org/10.1017/CBO9781107415324.004

Paneque, Rosa Jiménez and Habana (1998). Allá. Metodología de la InvestigaciónElementos Básicos para la Investigación Clínica. P. 1-95.

Pernía, J. and Fornés, J. (2009). Climate change in groundwater. Geociencia ambiental y de ingeniería, 17 (2), 172-178. https://doi.org/10.2113/gseegeosci.15.3.211

Promis, A., Caldentey, J. and Ibarra, M. (2010). Microclimate within a Nothofagus pumilio forest and the effects of regeneration logging Microclimate within a Nothofagus pumilio forest and the effects of regeneration logging. Forest, 31(2), 129-139. Retrieved from https://scielo.conicyt.cl/pdf/bosque/v31n2/art06.pdf

United Nations Development Programme (UNDP). (2006). More on scarcity: energy, poverty and the global water crisis. Human Development Report 2006. New York, USA.

United Nations Environment Programme (UNEP) (2008). Atlas of our Changing Environment. Nairobi, Kenya. Available at: na.unep.net/atlas.

UNDP (2007). Fighting climate change: solidarity in a divided world. Human Development Report 2007/2008. New York, United States.

Rosales, GO (2001). Evidence of climate change and climate variability in Ecuador.

Sánchez-Santilán, N., Sánchez-Trejo, R., Espino, G. and Garduño, R. (2014). Evolution of climate through Earth's history. Reflxiones Journal, 93(1), 121-132.

Serrano, S., Zuleta, D., Moscoso, V., Jácome, P., Palacios, E. and Villacís, M. (2017). Statistical analysis of monthly and daily meteorological data for the determination of climate variability and climate change in the Metropolitan District of Quito. La Granja, 16(2), 23. https://doi.org/10.17163/lgr.n16.2012.03

Sepúlveda, Yulian A. (2020). Importance of weather stations for decision making in Casadiego. 2020. No. May DOI 10.13140/RG.2.2.33323.46886.

Symborg (2024). Keys to helping crops in extreme temperatures. https://symborg.com/es/actualidad/claves-para-ayudar-a-los-cultivos-ante-las-temperaturas-extremas/ 15-06-2024

Sophocleus, M. (2002). Groundwater-surface water interactions: the state of the science. Journal of Hydrogeology, 10 (1), 52-67. ttps://doi.org/10.1007/s10040-001-0170-8

Suárez, J. (2009). Landslide: Geotechnical analysis (University). Spain.

Useros, LJ (2012). Climate Change: Its Causes and Environmental Effects. Real Academia de Medicina y Cirugía de Valladolid, 50, 71-98.

Uribe, Escudero and Alberto, Victor (2017). Environmental impact of temperature, relative humidity and rainfall variability on papaya production between Casma, Pomabamba and Recuay, 2012-2015, in the face of climate change effects.

Vinet, Luc and Zhedanov, Alexei (2010). A "missing" family of classical orthogonal polynomials. 7 November 2010. DOI 10.1088/1751-8113/44/8/085201.

Villalpando, F (2024): How does climate change impact on agricultural production? ClimaproAgro. Climate information for agriculture. Mexico

Viñas MJ, Fox K and Jacobs P (2024). NASA analysis confirms that 2023 was the warmest year on record. NASA Goddard Space Flight Center. Headquarters, Washington. https://www.nasa.gov/es. 12 January 2024

Wikipedia (2013). *Relative humidity*. Free encyclopaedia. https://es.wikipedia.org/wiki/Humedad_relativa

VIII. ATTACHMENTS.

Annex 1: Data on climate variables

DADOS DE TEMPERATURAS MÁXIMA, MINIMA, MEDIA, HUMIDADE RELATIVA E PRECIPITAÇÃO REFERENTES A 2013, 2014, 2015, 2016, 2017 E 2018 – DADOS CLIMATICOS

	Meses	Ano 2013					Ano 2014					Ano 2015				
		Temperaturas (°c)			H.R	Precipitação	Temperaturas (°c)			H.R	Precipitação	Temperaturas (°c)			H.R	Precipitações
		Max	Min	Media	(%)	(mm)	Max	Min	Media	(%)	(mm)	Max	Min	Med	(%)	(mm)
01	Janeiro	31.9	20.8	26.2	55	55	36.9	14.8	25.3	51	41.9	38.3	17.3	26.2	98	87.5
02	Fevereiro	32.8	19.5	26.6		55						37.7	16.3	26.6	97	27.4
03	Março	33.4	19.8	25.9		57	32.4	17.9	23.5	99	74	36.9	16.4	25.5	99	96.5
04	Abril	35.4	10.7	23.9	35	0.6	32.1	15.7	22.7	98	24.8	34.6	14.7	24.8	95	6.6
05	Maio	32.9	8.5	21.6	35	4.6	32.4	10.2	21.8	93	1.9	33.7	10.6	22.4	73	0.0
06	Junho	32.7	6.9	19.4	34	0.6	30.2	5.9	18.9	76	0.0	31.5	2.1	18.2	69	0.0
07	Julho	30.9	7.5	19.5	34	0.0	31.2	5.9	18.9	71	52.3	32.5	7.3	19.7	63	0.0
08	Agosto	34.4	7.8	21.2	24	0.0	35.1	7.8	22.3	56	0.0	36.6	6.9	22.5	71	43.3
09	Setembro	37.8	8.2	25.6	27	6.6	38.3	10.9	25.6	89	1.6	38.2	12.1	26.3	63	0.0
10	Outubro	39.1	14.8	28.2	25	2.6	38.8	14.1	28.0	81	3.5	38.9	15.4	29.1	71	3.9
11	Novembro	38.2	16.4	27.1	24	59.5	36.3	17.7	25.8	97	48.3	40.5	16.5	28.7	93	35.7
12	Dezembro	35.2	17.8	25.1	42	132.7	36.5	16.4	25.2	99	115.1	38.1	18.9	26.6	99	145.6

	Meses	Ano 2016					Ano 2017					Ano 2018				
		Temperaturas (°c)			H.R	Precipitação	Temperaturas (°c)			H.R	Precipitação	Temperaturas (°c)			H.R	Precipitações
		Max	Min	Media	(%)	(mm)	Max	Min	Media	(%)	(mm)	Max	Min	Med	(%)	(mm)
01	Janeiro	36.6	15.9	25,2	57	70.7	35.5	16.5	25.2	54	83.5	36.6	18.0	25.7	58	106.0
02	Fevereiro	36.6	18.5	24,1	64	78.8	36.3	16.5	24.1	70	113.6	34.8	15.5	25.5	60	76.8
03	Março	36.5	15.5		65	96.5	32.9	17.8	23.7	73	172.1	34.4	16.8	23.8	74	135.1
04	Abril	35.2	14.9	25.1	52	21.6	34.7	14.6	23.4	62	17.0	33.0	14.0	23.1	72	250.2
05	Maio	33.9	9.6	22.1	36	0	33.2	10.8	21.8	47	0	32.3	10.8	21.9	51	0.1
06	Junho	31.8	8.1	19.9	34	0	31.8	8.4	19.6	44	0	31.1	6.1	18.3	42	0
07	Julho	31.9	5.3	18.7	28	0	31.9	7.5	19.4	37	0	31.3	7.5	18.3	39	0
08	Agosto	35.9	9.4	22.9	22	0	35.1	9.0	21.8	27	0	34.6	9.7	22.3	26	0
09	Setembro	38.4	10.9	25.8	19	0	38.3	13.6	26.1	27	0	38.3	8.5	25.3	20	0.1
10	Outubro	39.6	16.1	29.2	22	2.4	39.3	13.7	27.0	28	37.5	39.2	15.5	26.9	29	6.5
11	Novembro	39.6	19.5	28.3	42	16.5	37.6	15.4	27.1	39	48.6	39.6	16.0	28.7	32	9.9
12	Dezembro	36.3	18.4	26.2	58	153.6	38.0	17.2	26.3	55	98.7	39.6	16.3	28.7	35	13.9

INAMET
MINISTÉRIO DAS TELECOMUNICAÇÕES E TECNOLOGIAS DE INFORMAÇÃO
INSTITUTO NACIONAL DE METEOROLOGIA E GEOFISICA
DEPARTAMENTO PROVINCIAL DO CUNENE

DADOS DE PRECIPITAÇÃO, TEMPERATURAS MAXIMA, MINIMA, MEDIA E HUMIDADE RELATIVA REFERENTE AOS MESES DE JANEIRO, FEVEREIRO, MARCO E ABRIL DE 2020 - DADOS CLIMATICOS

MESES	TEMPERATURAS (°c)			H.R (%)	PRECIPITAÇÃO (mm)
	MAX	MIN	MED		
JANEIRO	35.8	18.0	25.0	69	91.8
FEVEREIRO	35.3	17.1	24.7	71	183.2
MARCO	34.7	14.7	24.7	71	63.8
ABRIL	34.9	13.6	24.1	62	4.5
MAIO	32.7	9.2	21.5	45	0.0
JUNHO	30.1	5.6	18.1	41	0.0
JULHO	27.8	10.4	18.8	36	0.0
AGOSTO	33,5	7,3	21,2	62	0,0
SETEMBRO	38,2	12,9	25,4	74	0,0
OUTUBRO	40,9	12,8	27,4	92	11,6
NOVEMBRO	38,4	16,8	27,5	95	24,3
DEZEMBRO	38,1	16,2	27,5	96	53,9

O Chefe de Departamento

António José Pereira

DADOS DE TEMPERATURAS MAXIMA, MINIMA, MEDIA, HUMIDADE RELATIVA E PRECIPITAÇÃO REFERENTE AOS MESES DE JANEIRO À DEZEMBRO DE 2021 – DADOS CLIMATICOS

MESES	TEMPERATURAS (°c)			H.R (%)	PRECIPITAÇÃO (mm)
	MAX	MIN	MED		
JANEIRO	34,6	16,2	24,8	98	85,7
FEVEREIRO	35,6	17,3	25,5	54	62,1
MARCO	36,3	16,4	24,8	65	168,7
ABRIL	33,9	14,4	23,8	53	9,5
MAIO	32,2	10,6	21,3	51	15,5
JUNHO	31.8	6.7	18.6	41	0.0
JULHO	29.7	6.4	18.3	34	0.0
AGOSTO	35.5	9.7	22.0	25	0.0
SETEMBRO	39.1	12.8	26.7	19	0.0
OUTUBRO	39.2	12.8	28.1	22	5.3
NOVEMBRO	39,4	15,9	27,6	37	31,2
DEZEMBRO	36,7	17,0	27,2	44	51,7

MINISTÉRIO DAS TELECOMUNICAÇÕES, TECNOLOGIA DE INFORMAÇÃO E COMUNICAÇÃO SOCIAL
INSTITUTO NACIONAL DE METEOROLOGIA E GEOFISICA
DEPARTAMENTO PROVINCIAL DE METEOROLOGIA DO CUNENE

MAPA DE TEMPERATURAS MÁXIMA E MINIMA, HUMIDADE RELATIVA E PRECIPITAÇÃO REFERENTE AO II TRIMESTRE DE 2021 – DADOS CLIMATICOS

Dias	Tª Máxima °c			Tª Mínima °c			Humidade Relativa %			Precipitação mm		
	Abril	Maio	Junho	Abril	Maio	Junho	Abril	Maio	Junho	Abril	Maio	Junho
01	28,9	32,2	29,8	19,4	18,5	10,9	78	59	37	1,2	0,0	0,0
02	29,7	21,6	28,9	18,8	18,4	13,1	75	91	37	0,0	14,9	0,0
03	30,9	28,5	27,2	20,8	16,0	10,9	59	68	34	0,0	0,0	0,0
04	31,5	29,0	26,5	14,0	16,5	8,2	54	69	35	0,0	0,0	0,0
05	31,9	28,0	26,1	19,6	17,8	9,3	56	66	38	0,0	0,0	0,0
06	33,0	30,9	26,5	18,9	16,3	9,8	62	65	43	0,0	0,6	0,0
07	32,5	30,8	27,8	19,5	19,7	10,8	57	66	48	3,0	0,0	0,0
08	31,2	31,2	30,1	18,8	19,7	10,8	54	62	45	0,0	0,0	0,0
09	30,6	30,9	30,0	19,1	17,3	11,4	47	51	45	0,0	0,0	0,0
10	30,7	29,5	29,7	17,4	16,7	10,5	47	41	44	0,0	0,0	0,0
11	30,2	29,2	28,4	17,4	11,6	10,7	48	41	40	0,0	0,0	0,0
12	30,8	29,0	27,9	16,1	12,4	10,7	46	44	38	0,0	0,0	0,0
13	31,2	28,8	26,4	17,3	12,4	9,1	51	45	38	0,0	0,0	0,0
14	31,1	29,8	26,9	16,8	12,8	8,2	57	48	39	0,3	0,0	0,0
15	30,4	30,0	28,5	18,8	13,2	6,7	55	45	38	3,3	0,0	0,0
16	30,8	30,9	28,8	19,7	12,8	8,2	48	44	40	0,0	0,0	0,0
17	31,1	30,8	30,0	17,1	11,8	8,7	47	40	38	0,0	0,0	0,0
18	30,0	31,3	29,2	17,0	15,6	8,0	52	43	41	0,0	0,0	0,0
19	29,5	31,1	27,9	16,2	12,5	10,6	51	47	40	0,0	0,0	0,0
20	30,1	29,8	27,2	15,4	13,1	12,0	51	49	42	0,0	0,0	0,0
21	30,6	30,1	26,0	15,9	11,9	13,0	49	43	42	0,0	0,0	0,0
22	31,3	29,1	26,5	17,8	13,3	8,8	48	45	42	0,0	0,0	0,0
23	32,3	28,9	26,7	18,3	12,5	10,6	45	50	45	0.0	0,0	0,0
24	29,5	28,1	29,1	17,6	11,4	9,0	53	52	45	0,0	0,0	0,0
25	32,8	27,2	29,1	14,4	11,5	9,9	56	52	46	0,0	0,0	0,0
26	32,0	27,3	29,2	15,0	10,6	10,3	45	48	43	0,0	0,0	0,0
27	32,5	27,6	31,4	15,9	11,4	9,1	46	50	40	0,0	0,0	0,0
28	33,9	28,4	31,4	16,4	10,7	11,6	47	45	35	0,0	0,0	0,0
29	33,2	29,6	30,4	17,1	11,1	9,2	47	45	40	0,0	0,0	0,0
30	31,5	30,2	30,3	18,3	11,4	11,2	62	39	38	1,6	0,0	0,0
31		30,2			11,4			39			0,0	0,0
Total	632,0	911,4	853,9	525,0	545,6	301,3	1644	1592	1216	9,5	15,5	0,0
Média	31,2	29,4	28,4	17,5	17,6	10,0	55	51	40	9,5	15,5	0,0

MAPA DE TEMPERATURAS MÁXIMA E MINIMA, HUMIDADE RELATIVA E PRECIPITAÇÃO REFERENTE AO III TRIMESTRE DE 2021 – DADOS CLIMATICOS

Dias	Tª Máxima °c			Tª Mínima °c			Humidade Relativa %			Precipitação mm		
	Julho	Agosto	Set.	Julho	Agosto	Set.	Julho	Agosto	Set.	Julho	Agosto	Set.
01	29,7	27,1	34,9	8,9	10,9	14,7	37	35	12	0,0	0,0	0,0
02	29,6	29,9	35,4	9,9	10,5	12,7	33	36	15	0,0	0,0	0,0
03	29,4	30,9	34,6	8,8	12,5	13,3	33	34	21	0,0	0,0	0,0
04	29,0	31,8	34,6	7,5	11,1	14,7	35	33	22	0,0	0,0	0,0
05	28,9	31,7	34,6	6,8	10,8	15,4	37	29	21	0,0	0,0	0,0
06	29,9	31,5	35,9	7,6	11,9	15,5	34	27	14	0,0	0,0	0,0
07	28,4	31,2	35,5	9,4	11,2	15,7	30	27	15	0,0	0,0	0,0
08	28,0	32,5	36,1	9,8	9,9	18,0	38	27	19	0,0	0,0	0,0
09	26,8	31,9	35,3	9,5	10,0	18,5	38	26	24	0,0	0,0	0,0
10	29,2	32,5	34,6	10,3	11,6	21,1	36	25	24	0,0	0,0	0,0
11	28,1	32,4	33,3	9,5	11,5	19,6	37	25	23	0,0	0,0	0,0
12	28,3	32,9	33,7	9,5	12,1	18,3	33	25	24	0,0	0,0	0,0
13	28,2	33,2	35,1	9,1	10,1	15,3	32	26	22	0,0	0,0	0,0
14	24,8	33,4	35,2	9,9	10,3	17,2	36	24	20	0,0	0,0	0,0
15	27,1	31,8	35,8	9,9	13,4	16,2	32	27	19	0,0	0,0	0,0
16	28,6	31,9	35,2	11,2	13,2	15,5	36	33	18	0,0	0,0	0,0
17	28,0	32,9	34,9	10,5	13,5	15,2	43	26	16	0,0	0,0	0,0
18	27,1	33,2	35,9	15,4	12,2	15,2	39	25	21	0,0	0,0	0,0
19	27,1	32,7	34,9	14,2	10,2	17,3	34	18	20	0,0	0,0	0,0
20	28,1	30,8	35,8	9,9	9,7	16,4	32	21	18	0.0	0,0	0,0
21	28,2	31,3	37,8	8,6	10,9	18,3	30	24	17	0,0	0,0	0,0
22	27,0	32,6	38,0	8,7	10,8	21,8	29	21	19	0,0	0,0	0,0
23	27,0	31,0	39,1	8,3	12,6	24,3	22	22	15	0,0	0,0	0,0
24	26,5	33,4	38,2	6,4	17,9	19,7	22	23	17	0,0	0,0	0,0
25	27,1	33,9	37,4	11,1	13,9	17,4	36	21	23	0,0	0,0	0,0
26	25,6	35,2	37,7	12,3	15,7	20.9	38	22	16	0,0	0,0	0,0
27	25,6	32,5	36,9	8,0	16,5	16,2	37	20	21	0,0	0,0	0,0
28	27,0	30,9	34,2	7,6	12,0	16,9	34	27	28	0,0	0,0	0,0
29	27,2	29,7	33,9	7,8	13,4	17,3	32	18	23	0,0	0,0	0,0
30	26,9	30,4	36,1	9,5	11,9	15,0	32	15	16	0,0	0,0	0,0
31	26,9	35,5		9,1	12,3		32	13		0,0	0,0	0,0
Total	858,7	995,1	10,706	294,5	375,1	4,981	1054	775	569	0	0	0
Média	27,7	32,1	35,7	9,5	12,1	16,6	34	25	19	0	0	0

MAPA DE TEMPERATURAS MÁXIMA E MINIMA, HUMIDADE RELATIVA E PRECIPITAÇÃO REFERENTE AO IV TRIMESTRE DE 2021 – DADOS CLIMATICOS

Dias	Tª Máxima °c			Tª Mínima °c			Humidade Relativa %			Precipitação mm		
	Out	Nov	Dez	Out	Nov	Dez	Out	Nov	Dez	Out	Nov	Dez
01	33,9	36,2	35,9	17,5	17,4	21,9	15	18	41	0,0	0,0	0,4
02	33,3	37,3	32,6	14,2	19,2	19,2	14	17	58	0,0	0,0	3,8
03	32,9	36,4	32,6	12,8	17,7	21,3	15	17	58	0,0	0,0	2,2
04	34,1	39,4	34,1	16,5	17,9	20,4	24	20	62	0,0	0,0	4,2
05	35,6	38,3	33,2	16,0	23,5	19,8	16	25	64	0,0	12,8	0,0
06	37,7	37,4	34,6	17,5	23,0	19,2	17	49	50	0,0	5,0	1,0
07	35,8	36,1	35,4	20,4	19,4	20,0	36	59	34	0,0	1,7	0,0
08	37,9	35,4	35,9	22,1	19,5	21,5	24	51	46	0,0	0,0	0,0
09	37,7	34,0	34,9	21,7	22,4	22,8	20	40	49	0,0	0,0	0,0
10	39,0	34,4	32,7	20,8	23,3	18,9	13	43	62	0,0	0,0	1,5
11	39,1	33,4	32,2	22,2	23,4	22,4	10	44	60	0,0	0,2	19,0
12	37,8	34,7	31,4	19,9	19,9	18,8	24	50	70	0,0	0,4	6,1
13	37,2	33,7	30,7	19,6	21,7	19,8	21	47	78	0,0	0,1	8,3
14	38,0	36,0	32,1	19,4	22,6	20,9	18	53	63	0,0	2,6	0,0
15	38,2	30,3	34,7	22,9	19,5	22,3	20	71	53	0,0	0,1	0,0
16	35,6	36,2	34,3	17,1	19,2	20,1	19	57	48	0,0	1,3	0,0
17	35,7	34,9	34,2	17,2	20,7	21,1	15	48	37	0,0	2,3	0,0
18	36,8	30,0	34,3	17,3	19,8	18,2	18	64	32	0,0	1,2	0,0
19	38,1	35,0	36,2	23,8	19,1	20,6	15	41	31	0,0	0,0	0,0
20	33,7	37,0	36,7	22,6	22,6	20,4	31	26	34	0,0	0,0	0,0
21	34,9	35,9	35,6	19,3	18,8	21,6	38	23	44	0,0	0,0	2,0
22	34,7	34,7	34,8	16,8	18,1	18,3	21	24	49	0,0	0,0	3,2
23	35,9	34,6	34,7	16,1	15,9	21,1	20	23	30	0,0	0,0	0,0
24	36,9	39,5	35,7	18,2	18,3	18,2	16	20	26	0,0	0,0	0,0
25	36,4	36,8	36,0	20,5	18,9	17,0	16	21	25	0,0	0,0	0,0
26	36,3	35,6	36,5	17,2	18,1	18,7	17	20	24	0,0	0,0	0,0
27	35,7	36,9	36,1	23,9	19,2	20,9	23	23	23	0,0	0,0	0,0
28	33,3	35,8	34,9	14,7	22,5	17,1	40	38	23	0,0	2,9	0,0
29	36,7	34,1	35,8	22,3	16,1	17,4	45	47	28	4,6	0,0	0,0
30	36,0	35,9	35,9	20,5	21,0	21,7	42	46	36	0,7	0,6	0,0
31	36,6		35,1	21,6		20,6	16		40	0,0		0,0
Total	1122,2	1062,0	1072,6	592,1	597,0	933,1	682	1110	1256	5,3	31,2	51,7
Média	36,2	35,4	34,6	19,1	19,9	20,1	22	37	41	5,3	31,2	51,7

O Chefe de Departamento,

António José Pereira

MAPA DE TEMPERATURAS MÁXIMA E MINIMA, HUMIDADE RELATIVA E PRECIPITAÇÃO REFERENTE AO I TRIMESTRE DE 2022 – DADOS CLIMATICOS

Dias	Tª Máxima °c			Tª Minima °c			Humidade Relativa %			Precipitação mm		
	Jan	Fev	Marc	Jan	Fev	Marc	Jan	Fev	Marc	Jan	Fev	Marc
01	37,7	33,7	30,1	27,4	22,5	18,5	20	34	76	0,0	0,0	13,4
02	38,5	34,7	29,3	22,6	18,5	19,5	26	31	79	0,0	3,7	0,2
03	36,2	34,1	30,2	21,4	21,8	20,9	36	37	78	0,0	0,8	0,0
04	36,2	34,0	29,6	23,4	23,2	16,0	26	54	50	0,0	0,0	0,0
05	36,8	26,6	30,1	20,7	19,2	17,0	27	81	55	0,0	8,7	0,0
06	37,2	27,5	32,1	21,4	19,8	16,7	19	75	55	0,0	0,7	0,0
07	34,0	29,2	32,7	22,8	19,9	17,6	42	80	60	0,0	2,3	0,0
08	37,3	28,9	32,8	22,0	19,4	19,4	47	78	62	1,0	20,4	0,0
09	34,0	29,7	34,2	20,7	20,3	20,3	39	74	56	0,0	17,6	0,0
10	35,6	26,9	31,9	21,4	20,3	20,7	20	86	70	0,0	11,1	3,6
11	35,4	30,8	30,9	18,4	19,3	20,6	18	74	79	0,0	0,0	0,7
12	37,2	32,9	29,2	17,4	20,0	21,3	17	73	86	0,0	0,0	10,8
13	36,6	31,0	25,9	19,5	19,9	19,2	13	68	99	0,0	0,4	89,0
14	33,1	24,6	26,5	21,2	20,2	18,7	17	85	85	0,0	16,9	3,8
15	32,8	26,4	29,1	22,3	18,1	19,3	20	87	66	0,0	18,4	0,0
16	33,7	29,2	30,4	22,3	21,1	17,4	28	82	64	0,0	0,9	0,0
17	33,7	31,5	31,2	21,1	18,9	16,9	26	70	57	0,0	0,0	0,0
18	35,3	31,3	31,0	24,5	19,9	17,1	36	74	61	0,0	0,0	0,0
19	31,3	30,7	31,3	22,0	19,8	18,8	60	76	64	3,3	0,1	0,0
20	28,1	32,1	31,5	18,5	20,2	20,4	75	77	67	7,3	1,7	0,0
21	25,9	32,6	29,5	18,5	20,7	20,4	84	72	82	54,8	3,3	3,8
22	30,2	32,6	24,8	19,3	20,8	20,5	75	71	92	0,1	0,0	8,6
23	32,6	30,7	29,9	19,3	21,1	19,5	50	81	81	0,0	2,2	37,1
24	32,5	31,9	26,5	19,4	19,6	17,3	47	76	89	0,0	0,2	15,6
25	34,4	27,5	29,7	19,3	21,5	19,5	39	85	79	0,0	14,8	0,0
26	33,0	33,2	30,9	19,5	19,0	17,8	47	72	63	0,5	0,4	0,0
27	34,3	32,7	32,1	20,4	21,2	16,3	51	61	57	0,0	0,0	0,0
28	33,3	28,5	33,9	22,0	19,2	17,3	49	76	50	0,0	5,5	0,0
29	34,1		32,7	22,0		15,7	39		54	0,0		0,0
30	29,9		33,0	22,0		20,7	52		61	0,0		0,0
31	31,4		31,5	21,6		19,9	40		68	0,0		0,0
Total	1053,3	855,5	944,5	654,3	565,4	581,2	1203	1990	2227	67,6	126,4	186,7
Média	33,9	30,6	30,5	21,1	20,2	18,7	39	71	72	67,6	126,4	186,7

MAPA DE TEMPERATURAS MÁXIMA E MINIMA, HUMIDADE RELATIVA E PRECIPITAÇÃO REFERE NTE AO II TRIMESTRE DE 2022 – DADOS CLIMATICOS

Dias	Tª Máxima °c			Tª Mínima °c			Humidade Relativa %			Precipitação mm		
	Abril	Maio	Junh	Abril	Maio	Junh	Abril	Maio	Junh	Abril	Maio	
01	31,7	31,3	25,9	19,9	15,9	11,8	71	49	28	0,0	0,0	0
02	26,8	30,0	25,4	19,9	14,3	10,4	79	50	34	0,2	0,0	0
03	28,0	28,8	25,0	18,2	16,2	8,5	78	50	44	1,0	0,0	0
04	30,6	29,1	24,7	18,2	12,9	10,4	74	45	43	0,0	0,0	0
05	30,6	28,5	25,3	18,2	12,2	9,4	66	45	44	0,0	0,0	0
06	33,4	29,9	25,1	15,9	13,4	8,3	65	44	44	0,0	0,0	0
07	34,2	30,7	24,9	18,3	12,2	8,8	58	42	45	0,0	0,0	0
08	34,1	31,1	25,5	17,8	12,3	9,7	56	41	44	0,0	0,0	0
09	34,5	31,6	26,5	19,0	13,5	9,4	55	39	38	0,0	0,0	0
10	33,8	31,2	26,4	17,5	13,4	9,0	54	40	41	0,0	0,0	0
11	31,5	30,9	26,8	18,6	12,4	10,0	63	37	46	0,2	0,0	0
12	26,5	30,6	27,1	19,3	10,2	10,1	74	38	42	0,1	0,0	0
13	31,3	30,1	27,8	16,6	11,9	10,9	65	39	39	3,0	0,0	0
14	30,9	31,6	27,4	17,3	11,5	8,8	70	40	37	6,4	0,0	0
15	32,3	31,0	28,9	18,6	15,5	6,7	65	39	39	0,1	0,0	0
16	30,7	29,5	28,5	18,7	15,1	10,1	65	40	42	0,0	0,0	0
17	30,9	28,5	30,6	18,3	14,1	12,2	71	39	43	0,1	0,0	0
18	31,2	28,9	30,8	21,7	11,3	9,7	69	40	42	0,0	0,0	0
19	30,7	29,7	29,8	18,8	10,8	10,7	63	43	34	0,0	0,0	0
20	29,3	28,1	24,6	15,5	13,2	8,1	52	42	45	0,0	0,0	0
21	28,7	26,5	24,0	14,8	7,8	3,5	52	24	37	0,0	0,0	0
22	28,6	27,4	26,2	13,5	6,4	3,1	53	27	36	0,0	0,0	0
23	30,1	28,1	29,1	14,1	7,4	3,9	51	31	33	0,0	0,0	0
24	30,9	28,4	30,0	13,6	9,5	6,3	49	39	30	0,0	0,0	0
25	32,3	29,7	29,7	14,8	7,1	7,1	47	43	32	0,0	0,0	0
26	33,4	30,0	28,2	19,6	8,7	10,9	41	44	39	0,0	0,0	0
27	32,9	30,2	24,6	19,2	9,7	12,9	45	42	38	0,0	0,0	0
28	32,3	30,3	24,3	15,8	9,6	8,9	50	42	44	0,0	0,0	0
29	32,3	30,1	23,7	16,8	10,6	11,1	52	39	47	0,0	0,0	0
30	32,9	30,3	25,5	15,6	12,3	13,5	49	35	43	0,0	0,0	0
31		27,5			13,5			35			0,0	
Total												
Média	31,2	30,3	26,9	17,3	18,7	8,9	60	40	27	11,1	0,0	0,0

MAPA DE TEMPERATURAS MÁXIMA E MINIMA, HUMIDADE RELATIVA E PRECIPITAÇÃO REFERENTE AO III TRIMESTRE DE 2022 – DADOS CLIMATICOS

Dias	Tª Máxima °c			Tª Mínima °c			Humidade Relativa %			Precipitação mm		
	Julho	Ag	Set	Julho	Ag	Set	Jul	Ag	Set	Julho	Ag	Set
01	27,4	28,9	33,7	9,5	7,8	13,7	46	29	25	0,0	0,0	0,0
02	27,8	25,8	32,3	10,8	3,3	16,6	42	25	23	0,0	0,0	0,0
03	28,1	29,5	32,3	9,1	5,9	14,2	41	24	22	0,0	0,0	0,0
04	28,3	30,2	33,3	9,1	7,7	13,3	36	25	24	0,0	0,0	0,0
05	28,7	32,5	33,9	9,8	14,8	14,6	39	31	21	0,0	0,0	0,0
06	28,4	32,3	35,1	10,5	14,6	16,0	41	31	18	0,0	0,0	0,0
07	28,5	32,2	36,9	10,5	12,6	17,8	39	30	18	0,0	0,0	0,0
08	28,2	33,3	36,1	10,7	12,4	16,8	40	30	20	0,0	0,0	0,0
09	28,8	33,8	35,7	10,8	11,3	19,2	38	27	17	0,0	0,0	0,0
10	28,4	33,2	35,6	11,1	12,6	16,1	37	23	17	0,0	0,0	0,0
11	29,4	33,5	35,1	10,4	11,2	17,9	35	21	17	0,0	0,0	0,0
12	28,0	33,2	35,1	10,8	11,5	19,6	37	19	19	0,0	0,0	0,0
13	27,3	33,7	34,5	10,3	10,8	20,7	39	19	15	0,0	0,0	0,0
14	26,7	31,0	33,9	10,4	9,8	17,9	37	21	18	0,0	0,0	0,0
15	26,4	32,5	34,5	9,9	9,4	20,7	34	16	20	0,0	0,0	0,0
16	26,6	33,4	34,1	10,1	12,1	17,9	33	18	19	0,0	0,0	0,0
17	26,6	33,7	35,1	11,8	9,4	16,1	32	18	19	0,0	0,0	0,0
18	26,6	32,9	35,7	9,1	9,1	16,2	35	20	23	0,0	0,0	0,0
19	27,4	32,2	31,9	9,8	9,7	16,2	32	19	19	0,0	0,0	0,0
20	28,1	26,9	32,6	11,4	12,6	13,7	29	18	16	0,0	0,0	0,0
21	28,4	28,5	32,9	10,6	8,8	14,9	29	14	15	0,0	0,0	0,0
22	28,9	30,9	35,7	11,4	9,6	16,0	31	21	12	0,0	0,0	0,0
23	30,6	32,4	36,7	13,1	7,9	16,0	34	26	16	0,0	0,0	0,0
24	30,9	32,3	37,1	11,7	14,6	20,9	32	32	11	0,0	0,0	0,0
25	31,3	31,7	34,4	11,2	16,0	23,0	32	27	16	0,0	0,0	0,0
26	31,8	33,3	33,5	16,8	13,7	22,4	27	27	12	0,0	0,0	0,0
27	30,7	34,1	33,0	12,0	12,9	20,7	31	24	12	0,0	0,0	0,0
28	30,7	34,4	37,4	11,9	14,1	18,7	30	21	15	0,0	0,0	0,0
29	29,8	34,9	37,7	10,7	15,0	22,5	30	20	16	0,0	0,0	0,0
30	30,1	33,9	36,9	11,2	13,7	24,1	26	24	19	0,0	0,0	0,0
31	28,6	34,1		7,9	13,8		27	25		0,0	0,0	
Total	887,5	995,2	1042,7	334,4	348,7	534,4				0,0	0,0	0,0
Média	28,6	32,1	34,7	10,7	11,2	17,8				0,0	0,0	0,0

MAPA DE TEMPERATURAS MÁXIMA E MINIMA, HUMIDADE RELATIVA E PRECIPITAÇÃO REFERENTE AO IV TRIMESTRE DE 2022 – DADOS CLIMATICOS

Dias	Tª Máxima °c			Tª Mínima °c			Humidade Relativa %			Precipitação mm		
	Out	Nov	Dez	Out	Nov	Dez	Out	Nov	Dez	Out	Nov	Dez
01	37,0	34,5	37,1	19,3	20,3	22,3	15	38	34	0,0	0,0	0,0
02	38,2	35,3	35,9	19,3	20,6	22,6	13	40	35	0,0	0,0	0,0
03	37,5	34,9	37,3	18,9	20,2	22,3	16	34	30	0,0	0,0	0,0
04		33,2	34,8	18,3	20,1	22,5	20	18	57	0,0	0,0	0,0
05	38,0	31,0	36,9		21,4	21,7	14	15	25	0,0	0,0	0,0
06	37,0	33,0	35,1	19,2	21,0	20,7	25	22	13	0,0	0,0	0,0
07	37,0	32,1	35,7	19,1	17,8	20,1	13	34	33	0,0	0,0	0,0
08	36,5	34,0	34,9	18,4	19,1	20,6	33	17	15	0,0	0,0	0,0
09	36,0	33,9	36,4	18,4	17,0	19,3	31	16	7	0,0	0,0	0,0
10	36,9	35,6	35,5	21,0	19,6	22,8	21	20	39	0,0	0,0	0,0
11	36,9	34,8	33,7	19,2	19,1	21,7	17	19	21	0,0	0,0	0,0
12	37,0	33,0	34,6	20,3	16,6	147	20	16	25	0,0	0,0	0,0
13	36,1	33,3	35,9	21,2	15,2	20,5		11	26	0,0	0,0	0,0
14	36,1	36,7	29,4	21,5	16,3	20,4		10	66	0,0	0,0	0,0
15	32,1	37,4	27,3	23,4	18,2	17,3	43	13	73	0,0	0,0	0,0
16	27,6	36,8	31,7	20,6	15,8	20,6	54	28	57	0,0	0,0	0,0
17	27,6	36,9	31,5	16,7	19,6	21,8	50	40	30	0,0	0,0	0,0
18	34,0	37,9	35,8	21,6	18,2	21,6	39	36	22	0,0	0,0	0,0
19	35,2	36,1	35,9	21,0	20,6	20,8	33	39	18	0,0	0,0	0,0
20	35,5	34,3	35,9	21,9	21,6	21,9	37	45	18	0,0	0,0	0,0
21	35,5	33,8	36,7	21,9	23,9	19,1		46	14	0,0	0,0	0,0
22	35,8	33,8	36,2		22,0	25,3		44	29	0,0	0,0	0,0
23	36,8	36,5	35,5	22,5	23,4	21,4	25	30	36	0,0	0,0	0,0
24	36,8	31,9	36,1	22,5	16,8	219	25	60	37	0,0	0,0	0,0
25	31,4	33,6	35,0		18,8	22,5		53	51	0,0	0,0	0,0
26	36,6	33,8	35,0		19,8	20,8		54	38	0,0	0,0	0,0
27	35,5	35,4	36,4	22,3	21,7	21,7	37	38	10	0,0	0,0	0,0
28	35,7	34,8	37,0		19,9	24,4	37	42	27	0,0	0,0	0,0
29	35,7	34,7	33,8	20,3	21,9	20,1	54	36	62	0,0	0,0	0,0
30	29,9	36,3	30,8	16,7	19,6	20,8	69	35	60	0,0	0,0	0,0
31	30,9		34,2	20,0		21,0	51		44	0,0	0,0	
Total										0,0	0,0	0,0
Média										0,0	0,0	0,0

MAPA DE TEMPERATURAS MÁXIMA E MINIMA, HUMIDADE RELATIVA E PRECIPITAÇÃO REFERENTE AO I TRIMESTRE DE 2023 – DADOS CLIMATICOS

Dias	Tª Máxima °c			Tª Mínima °c			Humidade Relativa %			Precipitação mm		
	Jan	Fev	Marc	Jan	Fev	Marc	Jan	Fev	Marc	Jan	Fev	Marc
01	30,3	29,2	33,1	19,3	20,0	19,7	70	61	51			
02	30,4	30,9	29,8	21,0	21,5	19,1	61	51	66			
03	23,7	30,1	32,8	19,8	21,3	19,8	67	67	44			
04	20,1	32,1	33,0	18,7	20,8	18,0	98	48	20			
05	29,8	31,4	34,0	19,0	22,2	18,4	40	30	22			
06	31,1	30,9	35,9	19,6	20,8	18,8	29	59	21			
07	33,7	26,7	35,9	20,8	20,9	21,4	31	58	35			
08	32,3	28,6	31,0	22,9	21,7	21,5	27	65	50			
09	31,3	31,7	34,7	21,2	19,4	22,6	30	43	30			
10	33,2	33,1	34,4	23,1	18,8	22,4	44	46	40			
11	28,9	33,1	32,0	21,2	20,0	20,8	75	31	78			
12	31,6	33,8	32,8	21,6	20,3	22,6	67	31	68			
13	30,7	35,1	34,8	19,9	21,7	21,0	62	35	40			
14	31,1	35,6	35,0	18,0	23,4	21,2	68	31	39			
15	31,9	35,2	34,3	18,0	21,6	21,1	50	22	33			
16	29,6	33,3	33,4	20,7	18,2	20,2	71	27	47			
17	26,4	32,9	31,7	20,0	16,9	17,7	67	31	40			
18	25,7	31,9	33,5	19,0	13,7	20,0	85	23	50			
19	25,1	33,0	33,3	18,1	15,0	19,4	96	22	46			
20	22,3	33,1	33,9	20,6	15,4	19,3	90	24	49			
21	27,2	34,2	32,6	18,6	14,8	20,2	78	22	55			
22	24,6	35,4	32,4	17,2	17,4	21,3	71	26	47			
23	27,4	34,0	29,8	19,4	21,3	18,8	79	39	72			
24	28,4	32,9	32,0	18,5	20,8	20,5	79	48	53			
25	30,2	34,5	30,8	20,6	20,5	18,5	68	49	46			
26	31,0	34,3	25,6	19,0	20,2	21,2	71	46	84			
27	31,4	32,5	30,6	20,2	21,6	18,3	70	64	55			
28	30,1	27,9	28,4	21,1	19,7	18,6	68	-	77			
29	30,3		31,0	20,1		20,3	59		66			
30	27,8		31,8	17,9		20,6	77		63			
31	27,4			19,6		18,9	76		62			
Total	901,6	906,3	1006,6	548,8	548,8	622,2		1099,0	1565,0			
Média	28.6	30,2	32,5	19,7	19,6	20,1	65	78	51			

MAPA DE TEMPERATURAS MÁXIMA E MINIMA, HUMIDADE RELATIVA E PRECIPITAÇÃO REFERENTE AO II TRIMESTRE DE 2023 – DADOS CLIMATICOS

Dias	Tª Máxima °c			Tª Mínima °c			Humidade Relativa %			Precipitação mm		
	Abril	Maio	Junho	Abril	Maio	Junho	Abril	Maio	Junho	Abril	Maio	Junho
01	31,1	33,0	31,5	20,2	16,7	13,8	61	30	25			
02	32,0	32,5	32,1	19,4	15,1	15,5	59	27	29			
03	32,2	33,0	31,3	20,6	14,4	13,0	50	30	27			
04	31,8	32,5	32,7	20,6	14,4	12,4	55	31	22			
05	31,2	32,3	31,8	18,8	16,5	11,4	43	30	21			
06	32,6	32,6	31,8	17,8	16,5	11,4	54	25	13			
07	33,8	33,1	31,6	18,0	15,2	13,4	45	31	17			
08	33,4	32,8	30,9	19,6	17,8	12,6	39	35	21			
09	32,6	32,8	30,4	20,3	16,3	13,4	42	40	19			
10	33,6	33,5	31,2	21,3	16,3	13,4	46	24	20			
11	33,9	32,8	30,7	19,5	18,3	11,4	34	26	20			
12	33,4	31,7	27,2	19,5	15,1	11,8	40	30	22			
13	34,4	31,8	29,2	18,0	15,9	7,8	39	37	17			
14	33,7	31,7	29,4	19,8	14,2	9,2	35	28	24			
15	33,9	31,5	29,4	18,5	15,0	9,3	37	21	21			
16	33,4	30,2	30,1	16,6	19,9	7,7	33	34	15			
17	32,6	30,3	30,1	17,7	13,4	7,8	37	30	15			
18	32,7	30,5	30,0	16,7	11,2	9,2	45	24	18			
19	33,4	31,0	29,9	18,7	11,3	7,3	46	25	22			
20	31,6	32,3	30,0	20,2	12,3	8,1	45	18	22			
21	20,9	32,4	29,7	17,9	13,3	10,8	75	20	22			
22	29,2	33,1	29,6	18,0	15,6	10,8	72	25	20			
23	32,5	32,4	29,3	19,4	14,5	9,0	45	20	21			
24	32,7	32,4	28,3	20,6	16,4	11,2	46	28	22			
25	32,6	31,1	28,5	20,8	15,2	11,0	56	23	27			
26	33,3	31,7	28,5	19,1	13,4	9,5	48	21	34			
27	33,6		23,0	18,2	15,1	9,4	39	23	52			
28	33,5	31,7	22,4	18,2		3,3	28		32			
29	33,7	31,9	27,1	16,4	15,9	3,6	35	20	21			
30	33,4	29,8	28,2	16,0	12,6	6,8	33	23	27			
31		30,7			11,5			22				
Total	973,6	959,1	885,9	566,4	449,3	305,3	1362	801,0	661,0			
Média	32,4	31,9	29,5	18,9	14,9	10,1	45	27	22			

MAPA DE TEMPERATURAS MÁXIMA E MINIMA, HUMIDADE RELATIVA E PRECIPITAÇÃO REFERENTE AO III TRIMESTRE DE 2023 – DADOS CLIMATICOS

Dias	Tª Máxima °c			Tª Mínima °c			Humidade Relativa %			Precipitação mm		
	Julho	Agost.	Set.	Julho	Agost.	Set.	Julho	Agost.	Set.	Julho	Agos	Set.
01	28,4	26,2	33,9	13,8	9,3	13,8	21	15				
02	28,8	29,5	34,4	14,8	9,3	13,6	26	12				
03	29,0	30,8	35,7	11,3	8,8	14,2	22	31				
04	27,7	29,9	35,8	9,0	11,1	14,2	27	30				
05	27,4	29,0	35,6	9,1	11,9	14,3	35	24				
06	27,5	28,2	36,0	10,8	12,5	15,1	28	18				
07	27,8	28,5	35,9	12,2	12,9	15,2	20	24				
08	27,3	29,9	36,8	8,3	11,5	14,5	27	24				
09	27,8	30,8	36,9	8,2	12,8	16,6	27	25				
10	21,3	30,6	36,8	9,0	15,1	15,3	27	26				
11	21,3	32,1	31,5	8,3	14,5	17,0	15	25	23			
12	21,8	31,5	31,2	5,2	13,7	14,0	26	20				
13	24,8	32,0	34,2	6,7	16,0	14,0	35	10	14			
14	26,9	31,2	36,1	6,3	11,2	16,7		9	23			
15	29,3	33,0	37,0	7,1	11,3	17,7	21	20				
16	28,2	33,6	36,9	11,8	13,7	18,1	28	14	10			
17	26,6	33,4	37,5	9,6	12,5	15,5	28	20	10			
18	26,8	34,0	36,7	10,0	14,7	18,0	23	17	9			
19	27,6	33,2	34,6	8,4	12,6	17,2	22		24			
20	28,3	35,3	35,7	9,8	15,7	16,5	19	15	19			
21		35,7	35,5	12,0	13,2	18,7	18		30			
22	31,2	34,2	33,8	12,7	13,3	19,7	38		29			
23	31,4	34,4	35,3	17,1	14,4	18,7	32		27			
24	31,0	35,4	34,0	14,1	11,3	19,5	38		30			
25	29,5	32,9	31,5	9,8	13,0	14,8			17			
26	31,2	32,2	32,6		12,3	13,1	18	16	12			
27	32,0	30,0	34,4	12,0	11,4	14,8		13	13			
28	30,3	28,1	35,7	11,7	15,6	14,6		20				
29	26,8	30,7	37,8	10,5	19,5	19,5		41	20			
30	27,0	31,4	37,4	7,5	15,6	21,4		27	18			
31	26,5	32,9		10,2	14,7							
Total	862,6	980,7	1057,3	307,3	405,4	486,3	616	496				
Média	27,8	32,7	34,1	9,9	13,5	15,7	21	17				

MAPA DE TEMPERATURAS MÁXIMA E MINIMA, HUMIDADE RELATIVA E PRECIPITAÇÃO REFERENTE AO IV TRIMESTRE DE 2023 – DADOS CLIMATICOS

Dias	Tª Máxima °c			Tª Mínima °c			Humidade Relativa %			Precipitação mm		
	Out.	Nov.	Dez.	Out.	Nov.	Dez.	Out.	Nov.	Dez.	Out.	Nov.	Dez.
01	38,3		28,4	19,6		18,5	9					
02	39,4		28,3	20,9		20,5	14		75			
03	32,6		28,0	27,0		20,9	42		64			
04	37,5		30,9	19,2		19,9	16		63			
05	37,5		30,4	19,9		19,9	11		55			
06	37,5		30,2	21,9		17,5			67			
07			32,2			21,8			56			
08		38,7	32,4		18,3	21,7			60			
09		36,6	33,4		22,2	22,0			33			
10		37,7	32,7		23,9	21,6			39			
11		33,7	33,8		20,4	20,5		50	36			
12		35,7	34,1		23,5	20,8		34	41			
13		36,4	33,2		35,0	21,7		28	41			
14		37,0	32,2		25,2	22,0		25	26			
15		28,2	30,3		20,3	21,0		65	45			
16		33,8	34,7		20,9	20,8		52	37			
17		29,1	37,1		21,1	20,3		70	28			
18		32,8	32,9		23,8	23,2		40	69			
19		34,9	30,2		22,1	20,6		40	65			
20		28,7	32,7		20,9	23,1		83	56			
21		34,2	30,6		19,2	20,2		27	79			
22		35,7	30,4		25,2	21,3		23	62			
23		37,4	28,2		23,0	21,2		32	75			
24		31,9	28,6		20,2	21,1		62	77			
25		27,3	32,6		20,3	21,9		72	42			
26		27,3	31,6		18,1	19,1		68	40			
27		31,3	32,5		20,5	18,8		68	35			
28		33,7	32,2			18,3		49	44			
29		32,9	32,0		20,3	18,6		53	43			
30		30,1	32,5			20,0			47			
31			31,0			18,2			32			
Total	222,8	765,1	980,3	128,5	464,4	637,0						
Média	37,1	33,3	31,6	21,4	22,1	20,5						

O Chefe de Departamento

[illegible]

MIX
Papier aus verantwortungsvollen Quellen
Paper from responsible sources
FSC® C105338

Printed by Books on Demand GmbH, Norderstedt / Germany